GAS CHROMATOGRAPHY WITH GLASS CAPILLARY COLUMNS

SECOND EDITION

Gas Chromatography
with
Glass Capillary Columns

SECOND EDITION

WALTER JENNINGS

Department of Food Science and Technology
University of California
Davis, California

1980

ACADEMIC PRESS

A Subsidiary of Harcourt Brace Jovanovich, Publishers

New York   London   Toronto   Sydney   San Francisco

chem

ACADEMIC PRESS, INC.
111 Fifth Avenue, New York, New York 10003

United Kingdom Edition published by
ACADEMIC PRESS, INC. (LONDON) LTD.
24/28 Oval Road, London NW1   7DX

Library of Congress Cataloging in Publication Data

Jennings, Walter, Date.
    Gas chromatography with glass capillary columns.

    Includes bibliographical references and index.
    1. Gas chromatography.   2. Capillarity.   I. Title.
QD79.C45J46   1980   543'.0896   79–8851
ISBN 0–12–384360–X

PRINTED IN THE UNITED STATES OF AMERICA

80 81 82 83   9 8 7 6 5 4 3 2 1

CONTENTS

CHAPTER 3

Column Coating

CHAPTER 4

Inlet Systems

CHAPTER 5

Column Installation

CHAPTER 6

Measuring Column Efficiency

CHAPTER 7

Treatment of Retention Data

CHAPTER 8

**Temperature Programming
and Carrier Flow Considerations**

CHAPTER 13

Analysis of Materials of Restricted Volatility

CHAPTER 14

Instrument Conversion

CHAPTER 15

Applications of Glass Capillary Gas Chromatography

CHAPTER 16

Fault Diagnosis

APPENDIX I

Nomenclature

APPENDIX II

Liquid Phases

APPENDIX III

Porous Polymer Data

APPENDIX IV

Silylation and Derivatization Reactions

PREFACE TO SECOND EDITION

The need for this second edition is dictated by developments that have occurred in this rapidly changing field and by sins of omission and sins of commission in the first edition. Projections available at the time of this writing indicate that gas chromatography will remain the world's most widely used analytical technique for some time. It is applied to an ever-widening number of areas; a partial list would include the monitoring of contaminants in air, in food, and in water; quality control of petrochemicals; research in food flavors, essential oils, and fragrances; the detection of heavy metals in a variety of products; and diagnostics in clinical medicine. This catholicism of usage has resulted because gas chromatography offers (1) the best available method for the separation of volatile mixtures, (2) comparatively short analysis times, and (3) good sensitivity. These advantages have encouraged the development of methods permitting the conversion of nonvolatile compounds into volatile derivatives as an aid in their analyses.

Glass capillary gas chromatography with WCOT columns offers these same advantages, but to a greater degree than do packed or even SCOT or PLOT columns. Not only are WCOT glass capillary columns capable of yielding much more complete separations (i.e., higher plate numbers), but they can also provide much shorter analysis times and higher sensitivities. Additionally, they offer more inert systems that are capable of passing compounds that resist analysis in packed or metal columns (including nickel capillary columns). Higher coating efficiencies are possible on glass as compared to stainless steel or nickel capillaries, and localized defects that may occur after prolonged use (or misuse) can be visually diagnosed and frequently corrected.

As a result, more and more investigators are moving into the field of glass capillary chromatography. But while the rewards are great, the pitfalls are many. It is the author's hope that the efforts represented by this second edition will help those making this transition attain the former and avoid the latter.

PREFACE TO FIRST EDITION

Until quite recently, open tubular columns were covered by broad and strictly enforced patents. The lack of commercial competition added no impetus to the development of their full potential; most of those utilizing open tubular columns made their own, and glass open tubular columns were not commercially available. Later, a very restricted number could be obtained through one supplier in Switzerland, and another in Sweden. A major portion of the progress in glass open tubular columns, particularly in the area of application, has been made by individual scientists who had no choice but to become involved in glass capillary column technology in order to make these developments available to their individual research efforts. With the recent expiration of this restrictive patent, a number of suppliers have appeared who can now take advantage of these scattered individual efforts; a variety of glass capillary columns and associated paraphernalia have at last become available to the discriminating user. It is the goal of this book to serve as an introduction to glass capillary technology and as an aid in the selection, installation, evaluation, and use of glass open tubular columns.

CHAPTER 1

INTRODUCTION

1.1 General Considerations

"Chromatography" is a general term for separation processes in which the components of a mixture are repetitively equilibrated between two phases; normally one of these phases is fixed or stationary, and the other is mobile. When the mobile phase is a gas, either a liquid or a solid can be utilized as the stationary phase; these processes are most precisely termed "gas-liquid partition chromatography," and "gas–solid partition chromatography," respectively. The former, to which we will confine our attention, is generally termed simply "gas chromatography" or "GC." Occasionally one still sees the expression "GLC" (gas–liquid chromatography) or "VPC" (vapor-phase chromatography), but with a few exceptions these exist largely in the older literature. In the process of gas chromatography the stationary liquid phase is confined to a long tube, the column, in which it exists as a thin film that is either distributed over an "inert" granular support (packed columns) or supported as a thin coating on the inner surface of the column (wall-coated open tubular columns). The column, which begins at the inlet of the gas chromatograph and terminates at its detector, is adjusted to some suitable temperature and continuously swept with the mobile gas phase (carrier gas).

When a mixture of volatile components is introduced at the inlet, each constituent is swept toward the detector whenever it ventures into the moving stream of carrier gas. The molecules of those components that are more easily soluble in or exhibit stronger affinities for the stationary liquid phase venture into the carrier gas less frequently and require a longer period of time to reach the detector than do components that are less strongly oriented toward the liquid phase; hence separation is achieved.

Although our primary goal is an introduction to the practical considerations involved in the selection, installation, evaluation, and use of high-resolution open tubular glass capillary columns, at least a cursory knowledge of gas chromatographic theory is helpful in this regard. The symbols and nomenclature used throughout this discussion are detailed in Appendix I.

1.2   Theory of the Chromatographic Process

A compound subjected to the gas chromatographic process (a "solute") spends a fractional part of its transit time in the stationary liquid phase and the remainder in the mobile gas phase. Its equilibrium distribution between the two phases is reflected by the distribution constant K_D expressed as the ratio of the weights of solute in equal volumes of the liquid and gas phases:

$$K_D = \frac{\text{concentration per unit volume liquid phase}}{\text{concentration per unit volume gas phase}} \qquad (1.1)$$

K_D is a true equilibrium constant, and its magnitude is governed only by the compound, the liquid phase, and the temperature. Ester solutes would be expected to dissolve in, disperse through, and form intermolecular attractions with polyester-type liquid phases to a much greater degree than would hydrocarbon solutes exposed to this same liquid phase. Logically, the K_D value of an ester is higher than the K_D value of the hydrocarbon of corresponding chain length in a polyester liquid phase. As the temperature of the column is increased, both types of solute exhibit higher vapor pressures and their K_D values decrease, although those of the esters remain larger than those of the hydrocarbons. Among the members of a homologous series, of course, higher homologs possess lower vapor pressures and higher K_D values.

Ideally, the very short length of column occupied by a solute band on injection remains constant as the solute band traverses the column; as this tight, concentrated band leaves the column, it can then be delivered to the detector as a narrow sharp peak. In actuality, factors such as longitudinal diffusion that occur in both the gas and liquid phases cause broadening of the solute bands during the chromatographic process. The centers of the bands of solutes that have different K_D values will become increasingly separated as they progress through the column, but depending upon the degree of column efficiency, band broadening may cause the trailing edge of the faster component to interdiffuse with the leading edge of the slower component, resulting in incomplete separation and overlapping peaks (Figure 1.1). Hence the efficiency with which two components can be separated is governed not only by their relative retentions (*vide infra*), but also by the degree of band broadening that occurs. Insofar as the column is concerned, the separation efficiency is inversely related to the degree of band broadening; in a column of high efficiency, a minimum degree of band broadening occurs per unit of column length, and in a less efficient column, a higher degree of band broadening occurs per unit of column length.

Inasmuch as both are methods for separating mixtures of volatile compounds, it is not surprising that gas chromatography was promptly compared with the process of fractional distillation, and distillation terminology (i.e., "theoretical plates") was soon used to describe gas chromatographic separation efficiencies, albeit imperfectly (*vide infra*).

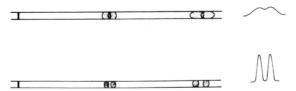

FIGURE 1.1 Band broadening as related to peak shape and component separation. The centers of the two bands are separated equally on both columns, but (top) a low-efficiency column with excessive band broadening causes interdiffusion of the separating components. As the components reach the end of the column, their concentrations in the carrier gas are relatively low and persist for a longer period of time, resulting in low, broad peaks; (bottom) a column of higher efficiency.

As already detailed, the separation efficiency of a gas chromatographic column is related to the degree to which a solute band
broadens (which is a function of the width of the peak, w) relative
to the length of time the band requires to traverse the column (its
retention t_R). The number of theoretical plates n is defined as

$$n = 16\left(\frac{t_R}{w_b}\right)^2 = 5.54\left(\frac{t_R}{w_{0.5}}\right)^2 \qquad (1.2)$$

where t_R is the time (or distance) from the point of injection
to the peak maximum, w_b the idealized width of the peak at the
baseline, and $w_{0.5}$ the width of the peak at idealized half height
(which should be at the point of inflection). The same units must
of course be used for the t_R and w measurements. Because the
precision with which w or $w_{0.5}$ is measured is critically important,
the use of an optical micrometer for this measurement is imper-
ative. It would be difficult to overemphasize the frequency with
which serious errors are made in this measurement. The measurement should be from the center of the rising recorder pen line
to the center of the pen line on the return stroke at peak half
height, assuming that the pen does not tip one way in its holder
on one stroke and the other way on the next. Results are more
consistent when the chart is placed on a lighted surface such as
an x-ray viewer while the measurement is made with a magnifying optical micrometer. Usually one measures from the outside
of one pen mark to the inside of the next; because the pen marks,
under magnification, frequently display a ragged edge and are
not always of uniform width, some degree of interpretation may
still be required. For extremely narrow peaks, it may be necessary
to measure the outside or overall distance and subtract the average
width of the pen stroke. Amazingly enough, those errors that still
occur most frequently favor the reporting investigator.

Obviously, longitudinal diffusion (and hence, gaseous dilution)
of components in the column is a major factor affecting band
broadening; smaller molecular weight components (usually characterized by smaller K_D values) would be expected to diffuse to
a greater degree per unit time than larger components. Diffusivity
would also be influenced by the density of the carrier gas. Consequently, the value of n is influenced not only by column effi-

ciency but also by the temperature, the test compound chosen, the type of carrier gas, and the degree to which the gas is compressed (*vide infra*).

A certain volume of carrier gas is required to conduct even nonabsorbed components through the column. This gas hold-up volume is given the symbol t_M and is equivalent to the volume of carrier gas required to conduct a nonsorbed component such as air through the column. Under steady-state conditions retention volumes are proportional to retention times, which, because they are more easily measured, and more generally used.

The more sensitive detectors required by most open tubular columns are not sensitive to air, so this value is usually estimated from the leading edge of a peak produced by methane injection, which at reasonable temperatures exhibits a vanishingly small K_D value. It is also possible to calculate t_M from the retentions of three members of a homologous series (see Section 7.2). Obviously, t_M contributes nothing to the separation process; indeed, simply by attaching a long empty fine-bore tube to the front end of the column, one could achieve very large values of t_M, leading to large values for t_R. This would give a grossly inflated figure for the number of theoretical plates n possessed by the column. More realistically, we deal with an adjusted retention time t_R':

$$t_R' = t_R - t_M \tag{1.3}$$

This value is used in calculating the number of usable or *effective* theoretical plates N:

$$N = 16\left(\frac{t_R'}{w_b}\right)^2 = 5.54\left(\frac{t_R'}{w_{0.5}}\right)^2 \tag{1.4}$$

Some authorities prefer the term "effective plates" rather than "effective theoretical plates." All the plates are theoretical, but not all are effective; hence "effective theoretical plates" seems a precise and nonredundant differentiation from "theoretical plates." Figure 1.2 illustrates accepted methods for determining the number of theoretical and effective theoretical plates.

Longer columns (of identical efficiency per unit length) will possess more theoretical plates, although because of complicating factors such as the increased pressure drop (*vide infra*), the relationship is linear only at $\bar{u}^{opt}$ (see Chapter 8). Efficiencies are sometimes expressed as the number of theoretical plates per meter of

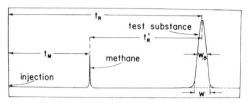

FIGURE 1.2 Alternate methods of determining the theoretical plate n and the effective theoretical plate number N; $n = 16(t_R/w)^2 = 5.54(t_R/w_{0.5})^2$; $N = 16(t_R'/w)^2 = 5.54(t_R'/w_{0.5})^2$, where $t_R' + t_M = t_R$, w is the idealized width of the test substance peak at baseline, and $w_{0.5}$ is the width of the test substance peak at half height. There are reports that the leading edge of the methane peak is a more accurate measure of t_M. (See Chapter 7 and Novák and Ruzickova [20].)

column length, e.g., n (or N) per meter. More often, however, the inverse value—the length of column occupied by one theoretical plate—is used. Once again, distillation terminology is employed, and this is termed the "height equivalent to a theoretical plate" (HETP). It is usually expressed in millimeters and is given the symbol h:

$$h = \frac{L}{n} \tag{1.5}$$

where L is the column length. The height equivalent to one *effective* theoretical plate (HEETP) is given the symbol H:

$$H = \frac{L}{N} \tag{1.6}$$

Obviously, smaller values of h (or H) indicate higher column efficiencies and greater powers of separation. The term h_{min} (or H_{min}) is used to express the value of h (or H) when the column is operating under optimum flow conditions (*vide infra*). Inasmuch as the values of n (and N) are affected by the column temperature, the test compound, and the nature of the carrier gas, these parameters obviously also affect h and H (*vide infra*).

As previously described, a solute spends a certain proportion of its time in the gas phase and the remainder in the liquid phase. The sum of these times is of course its observed retention time t_R. During the time that a substance is in the gas phrase, it is moving toward the detector at the same velocity as the carrier gas. Therefore, regardless of their retention times, all substances spend the same *length* of time, equal to t_M, in the gas phrase. Therefore the time spent in the liquid phrase will be equivalent to the adjusted retention time t_R'. The partition ratio (or capacity ratio) k is a measure of how long a time the molecules of a given species spend in the liquid phase relative to their time in the gas phase:

$$k = \frac{t_R'}{t_M} \tag{1.7}$$

It is useful to remember that

$$N = n\left(\frac{k}{k+1}\right)^2 \tag{1.8}$$

For solutes with very large partition ratios (i.e., large K_D and long retention),

$$\frac{k}{k+1} \simeq 1 \quad \text{and} \quad N \simeq n$$

Figure 1.3 illustrates the effect that the partition coefficient of the test compound has on the values of n and N for an actual experimental determination, and Figure 1.4 shows a graph of these data. Figures 1.5–1.8 explore the theoretical relationship between h_{min} (or H_{min}), the column radius, and the partition coefficient of the test compound. These graphs can be very useful in comparing

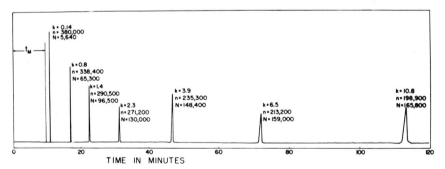

FIGURE 1.3   The effect of the partition ratio k of the test compound on the theoretical (n) and effective theoretical (N) plate number for an actual experimental determination. C_7–C_{13} n-paraffin hydrocarbons on an SE 30 glass WCOT capillary column at 130°C. Departure from the theoretical relationship shown in Eq. (1.8) is approximately 2%.

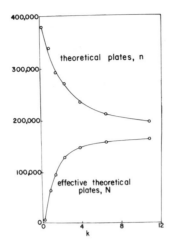

FIGURE 1.4   Theoretical (n) and effective theoretical (N) plate numbers as a function of the partition ratio k for a set of experimental data.

the different column test data presented by different manufacturers. See also Figures 8.3–8.5.

Logically, that proportion of the analysis time that a substance spends in the liquid phase, k, must relate to its distribution coefficient K_D. This relationship hinges on the relative availability of (i.e., the volumes of the column occupied by) the gas and liquid

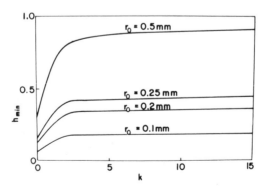

FIGURE 1.5   Relationship between the minimum height equivalent to a theoretical plate, h_{min}, and the partition ratio k of the test compounds for columns of several radii.

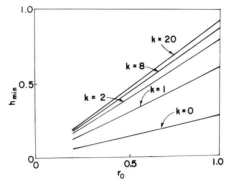

FIGURE 1.6   Relationship between the minimum height equivalent to a theoretical plate, h_{min}, and the radius of the column, r_0, for solutes with different partition ratios.

phases (i.e., the phase ratio) and is given the symbol β:

$$\beta = \frac{V_G}{V_L} \tag{1.9}$$

Referring back to Eq. (1.1), it can be seen that the distribution

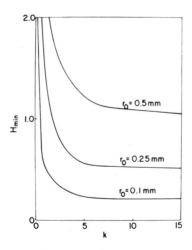

FIGURE 1.7   Relationship between the minimum height equivalent to an effective theoretical plate, H_{min}, and the partition ratio k of the test compounds for columns of several radii.

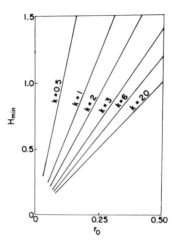

FIGURE 1.8 Relationship between the minimum height equivalent to an effective theoretical plate, H_{min}, and the radius of the column, r_0, for solutes with different partition ratios.

constant K_D can also be defined as

$$\frac{\text{amount in liquid/volume of liquid}}{\text{amount in gas/volume of gas}}$$

or

$$\frac{\text{amount in liquid}}{\text{amount in gas}} \times \frac{\text{volume of gas}}{\text{volume of liquid}}$$

The latter fraction has just been defined as the phase ratio β, and the former is the same as the partition ratio k. From this emerges a relationship we shall later utilize in rationalizing certain injection mechanisms and the interrelationships of several parameters on retention and separation characteristics.

$$K_D = \beta k \qquad (1.10)$$

It is apparent that β is a measure of the "openness of the column," and one would expect the phase ratios of open tubular columns to be appreciably larger than those for packed columns in which the packing not only limits the volume available for gas, but also increases the support area over which the liquid phase is distributed. Typically, packed columns have β values varying from perhaps 5 to 35, while in open tubular columns the values usually range from 50 to about 1500.

The inner surface area of the open tubular column (which at constant film thickness governs V_L) varies directly with common diameter, while the volume of the column (which governs V_G) varies directly with the square of the inner radius, i.e., the distance from the center of the column to the surface of the liquid phase coating. Hence both the diameter of the column and the thickness of the liquid phase film exercise an effect on the phase ratio of open tubular columns:

$$\beta = r_0/2d_f \qquad (1.11)$$

Figure 1.9 shows theoretical values calculated from this relationship.

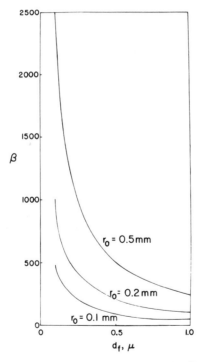

FIGURE 1.9 Phase ratios (β values) for columns of different radii, calculated from the relationship shown in Eq. (1.11).

Returning to Eq. (1.10), we have already seen that K_D, the product of βk, is a constant whose magnitude depends only on the solute, the nature of the liquid phase, and the column temperature; it is independent of parameters such as column diameter or the thickness of the liquid phase film. As β increases, k must decrease, and vice versa. Examined from another angle, we can argue that if the other variables are held constant, a larger diameter column will have a greater gas volume and, as V_G/V_L is greater, a larger β value. This larger gas volume will affect the retention behavior of each solute. K_D was defined as c_L/c_G [Eq. (1.1)]. The concentration terms c_L and c_G reflect the weight of solute per unit volume of gas or liquid phase. Columns with higher β values have higher gas/liquid phase ratios, and at any point in time more of each solute will be in the moving gas phase. Hence absolute retentions vary inversely with β; as the column diameter increases, d_f decreases and analysis times become shorter. Because all solutes are affected proportionately, relative retentions remain the same.

This relationship between retention behavior, separation efficiency, and the phase ratio β can be visualized rather simply in a qualitative manner. Consider two columns with the same liquid phase in which the second column has a larger diameter but the same film thickness, or a thinner film but the same column diameter, so that $\beta_2 > \beta_1$; the carrier gas flow rates are then adjusted so that $t_{M(1)} = t_{M(2)}$. If the columns are adjusted to the same temperature and the same solute is injected on both, $K_{D(1)} = K_{D(2)}$ (Section 1.2). From the relationship $K_D = \beta k$(10) and the given condition that $\beta_{(1)} < \beta_{(2)}$, it is apparent that $k_{(1)} > k_{(2)}$. Because $k = t_R'/t_M$ [Eq. (1.7)], $t_{R(1)}'/t_{M(1)} > t_{R(2)}'/t_{M(2)}$. Inasmuch as we have made $t_{M(1)} = t_{M(2)}$, $t_{R(1)}' > t_{R(2)}'$, and column (2) will exhibit shorter retentions and faster analysis times. At the same time, because solutes in column (2) spend less time in the liquid phase (t_R' is smaller), separation efficiencies will be lower.

The β value of a column can be determined by comparison with another column whose β value is known. If it is assumed that the known quantity of liquid phase is uniformly distributed on the inner periphery of open tubular columns coated by the static technique (a reasonable assumption, *vide infra*), the β value for that column can be calculated from Eq. (1.11). A test compound can then be chromatographed, and after measuring t_M, k for that compound can be calculated [Eq. (1.7)]. From Eq. (1.10), K_D for the test compound can then be determined, and as we have seen, this value will be the same for any column containing that same liquid phase at the same temperature. When the test compound is then chromatographed on the new column under the same conditions, β for the new column can be readily calculated from the relationship shown in Eq. (1.10).

1.3   SEPARATION OF COMPONENTS

As illustrated by Figure 1.10, the degree of component separation achieved is a function of

(1)   the ratio of their retention times, and
(2)   the sharpness of their peaks (or the number of theoretical plates possessed by the column).

The ratio of the corrected retention times of two components A and B is termed their relative retention:

$$\alpha_{A,B} = \frac{t'_{R(B)}}{t'_{R(A)}} = \frac{k_{(B)}}{k_{(A)}} \tag{1.12}$$

By convention, α is greater than unity. Solute pairs with large α value can be separated relatively easily even on low-resolution columns, but as this ratio approaches unity, columns with an increasingly larger number of theoretical plates are required to

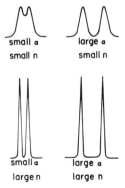

FIGURE 1.10   Relationship of column efficiency, as denoted by the theoretical plate number n, to the relative retention α in component separation.

achieve separation. Alternatively, of course, one can sometimes select another liquid phase in which the relative retention of those components is larger.

The degree of separation of two components is termed resolution R_s:

$$R_s = \frac{2(t_{R_A} - t_{R_B})}{w_{b_A} + w_{b_B}} \tag{1.13}$$

This relates resolution (or component separation) to the degree of peak broadening and retention time. As shown in Figure 1.11, a resolution of 1.0, while separating "idealized" peaks, actually results in a considerable degree of overlap. A resolution of 1.5 will usually achieve baseline separation, but tailing can cause complications. This is especially so when one is trying to resolve a minor component from a major one that tails as shown in the right-hand illustration of Figure 1.12.

If we assume that $w_1 = w_2$, then Eq. (1.13) can also take the

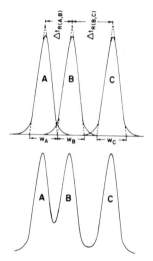

FIGURE 1.11 Separation of three components, A, B, and C; $w_A = w_B = w_C$; $\Delta t_{R(A,B)} = w_A$; $\Delta t_{R(B,C)} = 1.5w_C$. Top: solid line, individual contribution of each component to the detector signal. Dotted line, idealized peak shape. Bottom: resultant recorder trace. $R_{S(A,B)} = 1.0$. Note that considerable overlap occurs (bottom trace). $R_{S(B,C)} = 1.5$; overlap is very slight.

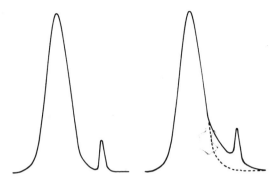

FIGURE 1.12 Separation complications caused by tailing. The purity of the minor component (as collected or delivered to a mass spectrometer) is much higher in the first case, where it follows a major component that produces a symmetrical peak, than in the second case, where the major component tails into the minor one.

form

$$R_s = \frac{n^{1/2}}{4}\left(\frac{\alpha - 1}{\alpha}\right)\left(\frac{k}{k + 1}\right) \qquad (1.14)$$

which emphasizes that resolution is proportional to the square root of the number of theoretical plates (or the column length) and that the partition ratio k also influences resolution. For very large values of k, $k/(k + 1) \simeq 1$, but for small values of k, this quotient is smaller. Hence better resolution should be achieved with later peaks (those having larger values of k) than with early peaks.

Another useful concept that relates to Eq. (1.14) is that from a knowledge of the α value for two compounds one can closely approximate the number of theoretical plates required for any desired degree of resolution:

$$N_{req} = 16R_s^2\left(\frac{\alpha}{\alpha - 1}\right)^2 \qquad (1.15)$$

Some supply houses sell columns on the basis of their theoretical plates n rather than effective theoretical plates N; by utilizing test compounds with relatively low k values, these efficiencies can be made to look even better (see Figures 1.5–1.8 and 8.3–8.5). When comparing columns from different suppliers using different systems, it is wise to remember that

$$N\left(\frac{k+1}{k}\right)^2 = n$$

Hence

$$n_{\text{req}} = 16R_s^2\left(\frac{\alpha}{\alpha-1}\right)^2\left(\frac{k+1}{k}\right)^2 \tag{1.16}$$

where k is the partition ratio of the second component.

A number of valuable concepts are embraced by the van Deemter equation [1], which permits evaluation of the relative importance of a series of parameters on column efficiency. In its abbreviated form, this is expressed as

$$h = A + \frac{B}{\bar{u}} + C\bar{u} \tag{1.17}$$

Where A includes packing and multiflowpath factors, B is the longitudinal diffusion term, C the resistance to mass transfer from the gas phase to the liquid phase and from the liquid phase to the gas phase (*vide infra*), and $\bar{u}$ the average linear velocity of the carrier gas. For a column of length L,

$$\bar{u} = \frac{L}{t_M} \tag{1.18}$$

Open tubular columns contain no packing and the A term becomes zero, reducing the van Deemter equation to a form known as the Golay equation [2]:

$$h = \frac{B}{\bar{u}} + C\bar{u} \tag{1.19}$$

Elimination of the A term from the van Deemter equation accounts for one of the major advantages of Golay or open tubular columns. Other factors have also been cited as at least partly responsible for the much higher efficiencies of open tubular columns: such columns also have much higher β values; much longer columns can be used before the pressure drop through the column becomes limiting; lacking points of contact between adjacent particles of coated solid support, the liquid phase has less tendency to depart from its form of a uniformly thin film. Another important advantage that is frequently overlooked relates to the fact that most packing materials are poor heat conductors. Particularly in a temperature-programmed mode, a range of temperatures must exist through every cross section of any packed column. The various molecules of each solute in that column must display a range of partition ratios, which would affect their retention behavior relative to one another and contribute to band broadening.

Obviously, our goal is the lowest possible value for h (or H), which is equivalent to the highest possible value of n (or N), commensurate with the degree of separation required, the efficiency of the column, the range of partition ratios embraced by the sample solutes, and the time available for the analysis (see Chapter 8). As h varies indirectly with the value of $\bar{u}$ in the B term and directly with the value of $\bar{u}$ in the C term, there must exist some optimum value of $\bar{u}$ at which the highest efficiency will be achieved for a solute of given partition ratio. This can be calculated, or determined graphically, and the interested reader is referred to more general or theoretical references [3–5]. Figure 1.13 shows a typical van Deemter curve.

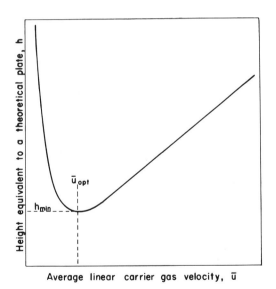

FIGURE 1.13   Typical van Deemter plot.

References

1. **van Deemter, J. J., Zuiderweg, F. J.,** and **Klinkenberg, A.,** *Chem. Sci.* **5,** 271 (1956).
2. **Golay, M. J. E.,** *in* "Gas Chromatography" (V. J. Coates, H. J. Noebels, and I. S. Fagerson, eds.), pp. 1–13. Academic Press, New York, 1958.
3. **Littlewood, A. B.,** "Gas Chromatography: Principles, Techniques, Applications," 2nd ed. Academic Press, New York, 1970.
4. **Kaiser, R.,** "Chromatographie in der Gasphase," 3rd ed., Vol. II. Bibliographisches Inst., Mannheim, 1975.
5. **Ettre, L. S.,** "Open Tubular Columns in Gas Chromatography." Plenum, New York, 1965.

CHAPTER 2

THE GLASS CAPILLARY COLUMN

2.1 Introduction

Ettre [1] pointed out that considerable confusion exists in the terminology that is used with open tubular columns and columns of capillary dimensions. Following his suggestion, we will use wall-coated open tubular (WCOT) to specify columns in which the liquid phase is deposited directly on the glass surface without the inclusion of any additive that might be considered solid support, porous-layer open tubular (PLOT) to describe columns in which the inner surface has been extended by substances such as fused silica [2] or elongated crystal deposits [3], and support-coated open tubular (SCOT) for columns in which the liquid phase is supported on a surface covered with some type of solid support material [4]. Ettre restricts the word "capillary" to uncoated tubing of capillary dimensions, but this latter suggestion has not been widely adopted.

The superior performance characteristics of glass as a column material have been recognized for a long time; the surface is more inert and less apt to react with substances being chromatographed, and defects in the column are more apparent and can be diagnosed. Consequently, many workers have long used glass-packed columns. In 1960 Desty *et al.* [5] suggested a simple design for a machine to draw and coil glass capillary tubing. Basically, a set of slow-speed rollers feeds a glass tube (6–10-mm diameter) into an electric furnace, and a second set of rollers operating at a higher speed draws the capillary out and feeds it into an electrically heated bending tube to form a continuous coiled length (Figure 2.1). Several commercial models are currently available; Figures 2.2 and 2.3 show two popular models.

With the ready availability of glass capillary tubing, a number of workers attempted its use as a more inert support material for open tubular columns. These early efforts met with a limited degree of success, and in the vast majority of cases the results were varied and lacked reproducibility. Most of these columns were of low quality and exhibited short lives, but those pioneers who did achieve results excited a great deal of interest [6, 7]. The chromatograms obtained by some of these workers demonstrated that glass open tubular columns were capable of a degree of resolution that had not been previously attainable. In addition, some compounds that failed to negotiate either packed or metal open tubular columns survived analysis in these more inert columns. This exciting prospect has led to a great deal of effort that continues to this day, as we try to gain a better understanding of the interrelated forces involved in coating glass capillary columns.

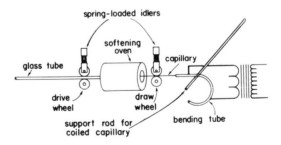

FIGURE 2.1 Schematic diagram of glass capillary drawing machine. (After Desty *et al.* [5].) As usually designed, the draw rate is fixed, and the feed rate, oven temperature, and bending tube temperature are variable. Increasing the feed rate or decreasing the oven temperature (within limits) produces a capillary of increased inside diameter. Too high a bending tube temperature produces a wavy capillary; too low a bending tube temperature leads to frequent breakage.

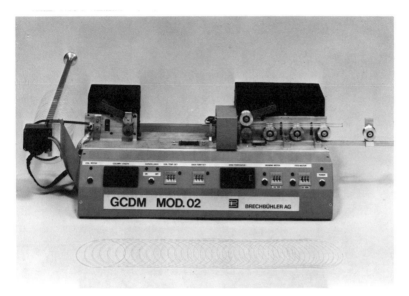

FIGURE 2.2   The Brechbühler glass drawing machine, designed for horizontal operation.

A clearer picture of these phenomena and how their interrelated effects are influenced by operating parameters and storage conditions could be of immense benefit and help us to achieve even higher efficiencies with more stable columns.

2.2   Surface Pretreatments

Surface chemists have long recognized that the degree of attraction between a solid surface and a liquid—wetting, if positive—is influenced by hysteresis, or the history of the solid surface. Glass, although not a solid in the true sense of the word, behaves in a similar manner. With a few notable exceptions, liquid phases casually deposited on untreated glass columns usually produce unsatisfactory coatings; the cohesive forces of the liquid are usually superior to the adhesive or wetting forces between the liquid and the glass surface, and the liquid phase draws up into discrete globules and beads (Figure 2.4). Such columns yield very unsatisfactory results because the concept of a thin

FIGURE 2.3   The Shimadzu glass capillary drawing machine, featuring vertical orientation and a low-amperage oven. ·

uniform film of liquid phase, essential to high-efficiency separations, has been violated.

This problem led to a number of suggestions for changing the wetting characteristics affecting the interior glass surface, either by the addition of surface active or wetting agents or by changing the characteristics of the surface itself. The wetting agent usually plays a dual role. When it functions in an ideal manner, it lowers the surface tension of the liquid phase and decreases the contact angle between the glass surface and the liquid phase; this encourages wetting and spreading. In some cases, it also satisfies most of the active adsorptive sites and may be regarded as an antitailing additive. Wetting agents that have seen wide usage include quaternary ammonium compounds [8], phosphonium compounds [9], nonionic detergents [10], and small addends of another more highly polar liquid phase. Temperature instability of these compounds, relative to that of the liquid phase, may be a problem in that they may result in lowering the upper temperature limit to which the column can be exposed. In addition, there is usually a small but recognizable effect on the retention characteristics of the column, which now behaves as it would with a mixed liquid phase [11]. As long as the levels of addend are low, this effect is usually minor.

The roughness, or rugosity, of a surface (actual area/theoretical area) plays an important role in its wettability by certain fluids. Microscopically roughened surfaces are sometimes easier to wet because of the change in contact angle relationships [12]. A reasonably fine finish stainless steel has a rugosity of about 1.4, whereas a fire-polished glass surface is nearer unity. Recognizing that an etched glass surface would have different wetting characteristics, several investigators explored surface pretreatments that included etching with reagents such as hydrochloric acid (usually dry vapors), sodium hydroxide, or hydrofluoric acid, sometimes at elevated temperatures. One of the most effective techniques is that described by Alexander and Rutten [13, 14]: The soda-lime glass capillary column is attached to a flask containing solid sodium chloride, and the addition of concentrated sulfuric acid generates gaseous hydrogen chloride at sufficient pressure (0.1–0.3 atm) to force the vapor through the column. Periodically, a few centimeters of column are cut off with a small torch and dropped, sealed end up, into a container of water. The HCl dissolves, drawing water into the small test section of cap-

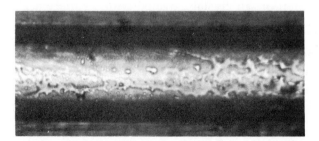

FIGURE 2.4 Photomicrograph of a badly deteriorated column.

illary to indicate the HCl/air ratio in the column. When at least 90% of the test section fills with water, both ends of the column are flame sealed, the column is then etched by subjecting it to elevated temperatures, and the residual vapors are finally flushed out with dry nitrogen.

Later work by these same authors [15] and by Badings et al. [16] indicated that the improved wetting characteristics of the etched surface are derived not only from the increased roughness or rugosity, but also and perhaps more probably from the leaching action of the etching agent, which leaves a residue of microscopic salt crystals on the glass surface.

Franken et al. [17] studied general methods of surface roughening and reported that when used with soda-lime glass, HCl etching proceeds readily but is a self-limiting process. The sodium chloride formed is initially present as globules, but on prolonged heating it assumes a cubic form. Krupcĭk et al. have also made an extensive study of the effects of etching and report that increased separation can be achieved when etched columns are used for the analysis of some types of compounds [18]. Onuska and Comba [19] used gaseous hydrogen fluoride as an etching agent on borosilicate capillaries, and found a whiskered surface resulted.

Glass possesses an extremely high surface energy, and a truly clean glass surface is almost never encountered [20]. More generally, adsorbed materials—usually water—modify this surface and dictate its behavior. The improved coating behavior of etched glass may result from substituting a salt-coated surface, which is more easily wetted by the liquid phase, for the hydrated glass surface normally encountered and which is less easily wetted [21, 22]. Instead of etching, some later workers wash the glass capillary with salt solutions. The column is then dried to leave a thin residual film of salt crystals, treated with a surface active agent, and then coated [23]. Deposits of microcrystalline sodium chloride, prepared by adding salt-saturated methyl alcohol to dichloromethane, have also been used to prepare a salt-roughened surface [24]. More recently, Badings et al. [25] suggested the use of amorphous silica, deposited by passing a dilute plug of water glass through the hydrogen fluoride etched capillary, followed by gaseous hydrogen chloride, as an alternate means of roughening the surface to encourage wetting.

Jennings [26] pointed out that the adsorbed water forming the hydration layer on glass was almost surely displaced during the drawing operation, but reformed as the drawn capillary cooled. He suggested that if the column could be continuously flushed with dry gas during drawing and then sealed until coated, the coating would be deposited on a dehydrated surface, producing a column whose liquid phase coating was more intimately bound and which should exhibit much greater stability. Simon and Szepesy [27] described a novel method of accomplishing this. A Teflon tube is inserted through the open end of the tube being drawn, and extended through the tube to the softened point within the furnace (a length of stainless steel capillary tubing might do even better). The stream of dry purge gas, introduced through this fixed central tube, turns back at this point and sweeps desorbed materials out of the open end of the glass tube being drawn. The authors report that the dehydrated capillary is completely wettable by nonpolar and most polar liquid phases and produces columns of increased stability.

Diez *et al.* [28] used a similar approach to flush the glass with a gaseous mixture of nitrogen and ammonia during drawing. They reported that when applied to borosilicate glass, the drawn tubing was less adsorptive and its rugosity was higher. The coated column gave superior results with polar compounds, including amines, pesticides, and drugs.

Grob [7] reported that nonpolar phases such as Apiezon could be coated on untreated glass drawn in the normal manner, but that polar phases were repelled. He suggested changing the interior glass surface by etching, and then filling the column with nitrogen saturated (at 0°C) with dichloromethane. The sealed column was then heated to a high temperature and the dichloromethane decomposed to produce a carbonized surface that was subsequently coated with liquid phase. Although Grob now uses a barium carbonate deposition (*vide infra*), some other workers continue to use the carbonization treatment with success. Nota *et al.* [29] reported that the graphite coating could be deposited in a simpler manner by dispersing colloidal graphite in dichloromethane with an ultrasonic generator and drawing the solution through the capillary with suction.

Schlute [30] reported that a thin layer of colloidal silicic acid on borosilicate glass capillaries produced columns of improved wetability that gave better separation of polar compounds. Grob fills the cleaned capillary with a solution of barium hydroxide, which, as it is expelled with carbon dioxide, forms a barium carbonate deposit on the surface [31–33]. The process reportedly produces a very stable column, although it is worth noting that Venema *et al.* [34] reported that the methyl silicone OV 101 suffered degradation when heatéd to 260°C in the presence of benzyltriphenyl-phosphonium chloride, barium carbonate, calcium chloride, sodium fluoride, and many other salts; soda-lime glass had a similar effect. Among the substances investigated, only borosilicate glass and sodium chloride appeared entirely inert toward the methyl silicone.

Several workers have studied the effects of silylation treatments on glass capillaries prior to coating [35, 36]. Novotny and Grohmann [37] found monochlorosilanes to be the most effective reagents and they used monochlorodimethyl- [3-(4-chloromethyl-phenyl)butyl]silane with both etched and previously untreated capillaries. Bartle and Novotny [38] reported on studies that correlated the wetting characteristics of various liquid phases with glass that had been subjected to selected silylation procedures, and some interesting evaluations of surface pretreatments in terms of column performance were presented by van Rijswick and Tesarik [39]. Grob *et al.* [40] have also recently suggested silylation as a means of column deactivation. In all these comparisons, however, it must be assumed that the untreated glass surface used for comparison was hydrated; treatments that would tend to remove the hydration layer may produce columns of improved performance and increased stability on untreated glass (*vide infra*).

Cronin [41] described another procedure that may be regarded as a surface pretreatment. The column was rinsed with a dilute solution of Carbowax 20 M in dichloromethane, dried, sealed under nitrogen, and heated to 280°C for 16 hr. After rinsing with dichloromethane, it was coated by a dynamic technique (*vide infra*). Somewhat similar treatments have been suggested by others [42–44].

Gordon *et al.* [45] coated the capillary first with a polymer, and then deposited the stationary phase on this intermediate layer. Some workers have emphasized that capillaries freshly drawn from tubing cleaned in alkaline permanganate seem to exhibit better coating characteristics [46, 47].

Other surface treatments, such as precoating with silicon dioxide [48] or creating a "whiskered" surface [49, 50], have been suggested, but such surfaces more nearly resemble those of PLOT than WCOT columns. While they would be expected to exhibit higher capacities, in that they contain more liquid phase per unit of column length and hence have smaller β values, a proportionate decrease in column efficiency is to be expected (see Chapter 1). Some argue that this generalization is incorrect, in that rather than being immersed in a thicker layer of liquid phase, the whiskers merely serve to increase the extent of surface coated with a normal film thickness. The whiskered surface is however highly active and requires extensive deactivation treatments before coating. N-cyclohexyl-3-azetidinal (CHAZ) has been used for column deactivation [50]. The crystalline material, m.p. 80°C, is unstable and is best stored at low temperature. For deactivation the column is coated dynamically with a 5% CHAZ solution, sealed, heated to 125°C overnight and rinsed thoroughly. The column begins to lose deactivation at temperatures of about 240°C. Other methods of column deactivation, applicable to the coated column, are discussed in Sections 5.2 and 5.5.

Most workers draw heavy-walled tubing, resulting in a heavy-walled capillary, which for a 0.25-mm i.d. may measure 0.75–1-mm o.d. Standard wall tubing results in a thinner-walled capillary; typical measurements are 0.25-mm i.d. and 0.5-mm o.d. Those that prefer the heavy-walled variety stress the increased strength and rigidity; those preferring the thinner wall claim that, because of the greater flexibility, these tubes are easier to work with and the column ends can be straightened more readily, so they can easily be extended far into inlets or detectors (*vide infra*). Both borosilicate and soda-lime glass have been used to prepare capillary columns. Two advantages have been claimed for the latter: It is more easily etched (most etched columns are based on soda-lime glass), and it is more flexible and is therefore less susceptible to breakage by oven air currents, etc. Advantages cited for borosilicate include the fact that a lower degree of adsorption or decomposition of silylated fatty acids (and presumably other silyl derivatives) is observed on silicone-coated borosilicate columns than on silicone-coated columns drawn from soda-lime glass [51–54]. On the other hand, de Nijs *et al.* demonstrated that endrine suffers extensive decomposition on borosilicate columns coated with methyl silicone [55]. The lower bleed rate of borosilicate methyl-silicone columns as compared to soda-lime methyl-silicone columns has been noted by several workers [53, 54].

Dandeneau and Zerenner [56] investigated capillaries drawn from various glasses, including an uranium glass and a potash-soda lead glass. While the former gave very poor performance, the lead glass produced columns that were, except for sulfur-containing compounds, much less active. They also investigated the use of fused silica as a column material. Figures 2.5–2.8 illustrate a variety of sample substances on several of the glasses investigated by Dandenau and Zerenner.

The words "quartz" and "silica" are occasionally used almost interchangeably, and some word on nomenclature is probably in order. Naturally occurring quartz (principally from Brazil) is processed commercially to produce a fused quartz. This material is available in ingots, bars, and tubing and is characterized by a high metallic oxide content, dominated by aluminum and iron oxides. Several refined grades of fused quartz are also commercially available. Although fused quartz is sometimes used to produce commercial columns, most utilize a synthetic fused silica, made from silicone tetrachloride.

The prospects of columns made from these materials have held a strong allure for many years, but a variety of difficulties slowed their development. By utilizing columns with exceedingly thin walls (0.20-mm i.d. × 0.25-mm o.d.) Dandeneau and Zerenner [56] produced an inherently straight but highly flexible column. Sophisticated drawing machinery, based on fiber optics technology and employing drawing temperatures of approximately 2000°C was employed in the production of the capillary. Later developments indicate that standard glass drawing machinery can be adapted to the drawing of fused quartz or fused silica capillaries by substituting oxy-acetylene flames for the normal softening oven in the drawing of heavy-wall (nonflexible) capillaries, and an electrically heated high temperature oven for thin-wall (flexible) capillaries.

The thin-wall columns are extremely fragile as drawn, and shatter on the slightest provocation. To guard against scratches in the outer wall which lead to fracture and shatter, the outside of the column must be immediately coated with a protective sheath as it emerges. In the early development of these columns silicone rubber was used for this purpose. A thinner polyimide coating

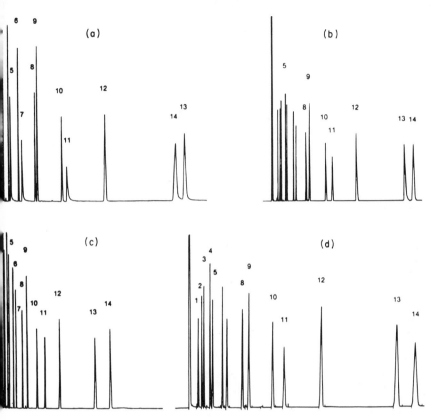

FIGURE 2.5 A test mixture on Carbowax 20 M coated columns of several different glasses: (a) uranium glass, (b) soda-lime glass, (c) potash soda lead glass, and (d) fused silica of 211 μ i.d. All chromatograms are isothermal at 140°C; except for the fused silica, all columns were 0.25-mm i.d. × 20–22 m, with an average carrier gas velocity of 25–30 cm/sec. The fused silica column had an inner diameter of 0.211 mm. Test substances were (1) *n*-hexanol, (2) nonanol, (3) 4-decanone, (4) pentadecane, (5) *n*-octanol, (6) hexadecane, (7) *n*-nonanol, (8) heptadecane, (9) naphthalene, (10) octadecane, (11) nicotine, (12) nonadecane, (13) eicosane, (14) 2,5-dimethylphenol. (From Dandeneau and Zerenner [56].)

is presently used by most manufacturers; this has excellent mechanical resistance and good temperature tolerance, but has a tendency to render the column opaque. This of course makes difficult the discernment of localized defects in the column; some progress is being made in the development of thermally stable transparent coatings.

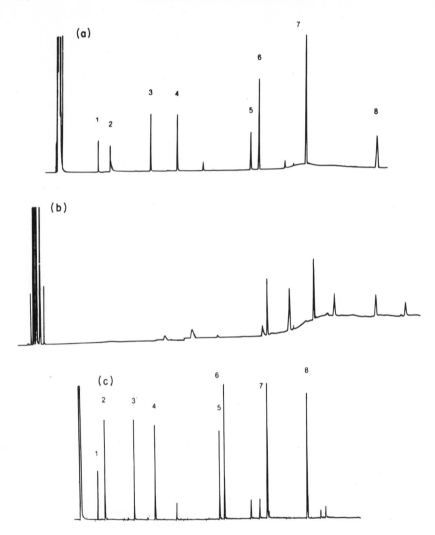

FIGURE 2.6 Mercaptans and sulfides on SP 2100 columns of (a) soda-lime glass, (b) potash soda lead glass, and (c) fused silica. Column dimensions as detailed in Figure 2.5; 50°C 2 min, programmed at 10°C/min to 190°C and held 5 min. Chart speed 2 cm/min. Injector temperature was maintained at 100°C to minimize thermal degradation of these labile compounds. Test substances were (1) 1-hexanethiol, (2) benzenethiol, (3) 1-octanethiol, (4) 1-nonanethiol, (5) phenyl sulfide, (6) 1-dodecanethiol, (7) benzyl sulfide, (8) benzyl disulfide. (From Dandeneau and Zerenner [56].)

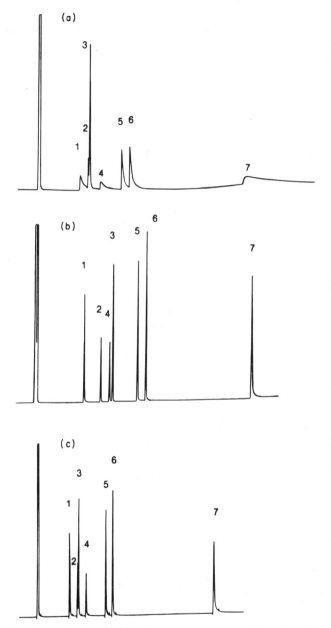

FIGURE 2.7 Phenols on SP 2100 columns of (a) soda-lime glass, (b) potash soda lead glass, and (c) fused silica. Column dimensions as detailed in Figure 2.5; 100°C 2 min, programmed to 190°C at 10°C/min and held 5 min. Chart speed 2 cm/min. Test substances were (1) phenol, (2) *o*-cresol, (3) 2,6-dimethyl phenol, (4) *m*-cresol, (5) 3,5-dimethyl phenol, (6) 3,4-dimethyl phenol, (7) 1-naphthol. (From Dandeneau and Zerenner [56].)

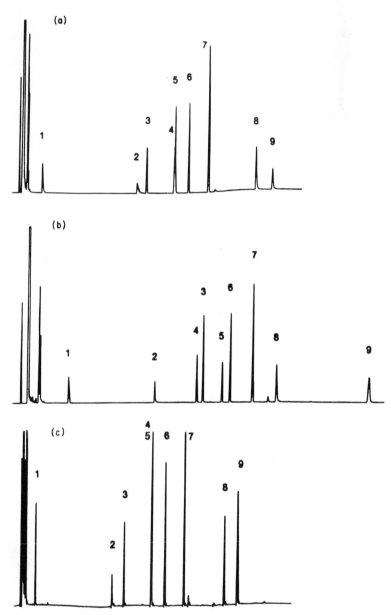

FIGURE 2.8 Amines on SP 2100 columns of (a) soda-lime glass, (b) potash soda lead glass, and (c) fused silica. Conditions same as in Figure 2.7. Test substances were (1) 2-ethylhexylamine, (2) *p*-phenyldiamine, (3) dicyclohexylamine, (4) 3,4-dichloroaniline, (5) *N,N*-dimethyldodecylamine, (6) *p*-nitroaniline, (7) diphenylamine, (8) 2,5-diethoxyaniline, (9) *n*-hexadecylamine.

The extreme flexibility of the thin-wall small bore fused silica (and fused quartz) columns can in some cases lead to minor problems in the area of column connection and on-column injection (Section 4.5); guiding the column into constricted passages and past hidden projections has been compared to threading a needle with an extended piece of limp thread. Larger diameter thin-wall columns can be forced into essentially the same bends, knots and convolutions without breaking, but they are somewhat stiffer, which some workers regard as a distinct advantage.

Both fused quartz and fused silica surfaces possess low surface energies, which make their wetting (and coating) characteristics different from those of conventional glasses. The more polar liquid phases are usually rejected by the untreated surface, and thicker films regularly lead to globule formation. The surface can be etched or eroded (rather than leached) to some degree, but because the flexible columns have such a restricted wall thickness, this is generally viewed as a high risk process. A variety of treatments—including silylation processes—are currently being explored by many workers in attempts to modify these surfaces so they will accept a wider range of liquid phases.

There are reports that silica columns catalyze the breakdown of most organic macromolecules and exhibit high bleed rates [33]. Polar liquid phases were found to exhibit lower temperature limits on silica columns, and apolar silicone liquid phases underwent a slow change from a liquid to a quasi-solid state [58]. The findings of Dandeneau et al. [57] are in sharp contrast to these reports, which may reflect differences in the grades or purities of the starting materials used for columns in the two cases. Results from this latter group indicate that their columns are in most cases more inert, possess excellent temperature stability and exhibit good efficiencies. Those active sites that they do possess may be due to strained siloxane bridges, formed when these glasses are subjected to temperatures in excess of 700°C [59]. At any rate, they do appear to be more inert, and at the time of this writing their only drawbacks would appear to be the restricted number of liquid phases that can be coated, and the lower degree of transparency, which makes the discernment and removal of localized defects more difficult than with conventional glass columns.

From this brief survey, it is evident that none of these materials—soda lime glass, borosilicate glass, lead glass, fused quartz, or fused silica—produces a surface that is completely inert, and that two general approaches have been used in efforts to better control their surface activities. One has involved covering the active surface (e.g., barium carbonate depositions) or changing the nature of the active groups (e.g., silylation, Carbowax 20 M); the other approach is to employ a substrate that lacks active sites (e.g., fused silica). Neither approach has yet been entirely successful, but there has been good progress on both fronts.

Both Schomburg [54, 60] and Grob and Grob [61] have stressed the importance of testing uncoated capillaries for activity, and have suggested systems to accomplish this. Venema *et al.* [34] have presented some results that need to be carefully evaluated if one is interested in producing stable silicone columns. Their results indicate that methyl silicone liquid phases heated on borosilicate glass or in contact with sodium chloride exhibit good stability, but that these liquid phases degrade when heated on soda-lime glass, or in contact with barium carbonate, sodium fluoride, calcium chloride, benzyltriphenylphosphonium chloride, and several other salts.

2.3   High-Capacity Systems

The major limitation of the glass capillary column is its restricted sample capacity. Capacities can be increased by resorting to larger diameter or thick-film columns, but one pays a price in reduced separation power. Other have used a whiskered surface as discussed previously, or SCOT, PLOT (Section 3.4), or micropacked (Section 3.6) columns; there are disadvantages associated with each of these approaches. Alternatively, high capacity smooth-wall open tubular systems have been investigated by several workers.

Janik [62] proposed coating the capillary spaces between a bundle of circular wires or plates to produce multicapillary columns for preparative scale GC. Pierce *et al.* derived a novel method of placing several capillaries within a larger tube and then drawing the entire bundle (Figure 2.9); while it would then be necessary to plug the intersticial spaces between the resultant capillaries, methods of accomplishing this were also proposed [63]. At our

FIGURE 2.9 Multicapillary column. (After Pierce *et al.* [63].) Magnification ×
100.

present state of the art, however, it is doubtful whether the phase
ratios would be identical in each flow path. This would result in
each solute exhibiting a range of partition ratios, resulting in
broadened peaks.

Several workers, (e.g., Papendick and Baudisch [64], Du Plessis
and Torline [65]) in efforts to reap both the benefits of the larger
surface areas that accompany wide-bore capillaries and the higher
efficiencies of small bore columns, have studied flattened capillaries. A variety of atypical column configurations, including hel-
ically coiled open tubular capillaries, were explored by Desty and
Douglas [66]. In recent discussions, Dandeneau [67] reported that
flattened capillaries had been prepared from both conventional
glass and thin-wall fused silica; the former exhibited a marked
increase in capacity and efficiency, but difficulties were encountered with coating the latter, but last reports indicate progress is
being made in this direction.

With this brief introduction to glass, let us now turn our attention to the coating process itself. Satisfying though it may be when the glass drawing machine, after long periods of frustration, finally begins spinning out acceptable lengths of glass capillary tubing, the battle has not yet begun. Given sufficient patience and resources, anyone can eventually pull glass capillary tubing. It is the ability to coat columns routinely with highly stable and uniformly thin, coherent films of liquid phase that makes the difference between success and failure in producing WCOT columns.

References

1. Ettre, L. S., "Open Tubular Columns; an Introduction," Publ. GCD-35. Perkin-Elmer Corp., Norwalk, Connecticut (1973).
2. Cronin, D. A., *J. Chromatogr.* 48, 406 (1970).
3. Schieke, J. D., Comins, N. R., and Pretorius, V., *J. Chromatogr.* 112, 97 (1975).
4. Nikelly, J. G., *Anal. Chem.* 45, 2280 (1973).
5. Desty, D. H., Haresnip, J. N., and Whyman, B. H. F., *Anal. Chem.* 32, 302 (1960).
6. Halász, I., and Horvath, C., *Anal. Chem.* 35, 499 (1963).
7. Grob, K., *Helv. Chim. Acta* 48, 1362 (1965).
8. Metcalfe, L. D., and Martin, R. J., *Anal. Chem.* 39, 1204 (1967).
9. Malec, E. J., *J. Chromatogr. Sci.* 9, 318 (1971).
10. Mon, T. R., Forrey, R. R., and Teranishi, R., *J. Gas Chromatogr.* 5, 497 (1967).
11. Jennings, W. G., *Chem., Mikrobiol., Technol. Lebensm.* 1, 9 (1971).
12. Wenzel, R. N., *Ind. Eng. Chem.* 28, 988 (1936).
13. Alexander, G., and Rutten, G. A. F. M., *Chromatographia* 6, 231 (1973).
14. Alexander, G., Garzó, G., and Pályi, G., *J. Chromatogr.* 91, 25 (1974).
15. Alexander, G., and Rutten, G. A. F. M., *J. Chromatogr.* 99, 81 (1974).
16. Badings, H. T., van der Pol, J. J. G., and Wassink, J. G., *Chromatographia* 8, 440 (1975).
17. Franken, J. J., Ruten, G. A. F. M., and Rijks, J. A., *J. Chromatogr.* 126, 117 (1976).
18. Krupcík, J., Kristin, M., Valichovicová, M., and Janiga, S., *J. Chromatogr.* 126, 147 (1976).
19. Onuska, F. I., and Comba, M. E., *J. Chromatogr.* 126, 133 (1976).
20. Olsen, D. A., and Osteraas, A. J., *J. Phys. Chem.* 68, 2730 (1964).
21. Fox, H. W., Hare, E. F., and Zisman, W. A., *J. Phys. Chem.* 59, 1097 (1955).
22. Hare, E. F., and Zisman, W. A., *J. Phys. Chem.* 59, 335 (1955).
23. Watanabe, C., and Tomita, H., *J. Chromatogr. Sci.* 13, 123 (1975).
24. de Nijs, R. C. M., Rutten, G. A. F. M., Franken, J. J., Dooper, R. P. M., and Rijks, J. A., *HRC&CC* 2, 447 (1979).
25. Badings, H. T., van der Pol, J. J. G., and Wassink, J. G., *HRC&CC* 2, 297 (1979).

26. Jennings, W. G., *Chromatographia* **8**, 690 (1975).
27. Simon, J., and Szepesy, L., *J. Chromatogr.* **119**, 495 (1976).
28. Diez, J. C., Dabrio, M. V., and Oteo, J. L., *J. Chromatogr. Sci.* **12**, 641 (1974).
29. Nota, G., Goretti, G. C., Armenante, M., and Marino, G., *J. Chromatogr.* **95**, 229 (1974).
30. Schulte, E., *Chromatographia* **9**, 315 (1976).
31. Grob, K., and Grob, G., *J. Chromatogr.* **125**, 471 (1976).
32. Grob, K., Jr., Grob, G., and Grob, K., *HRC&CC* **1**, 149 (1978).
33. Grob, K., Grob, G., and Grob, K., Jr., *Chromatographia* **10**, 181 (1977).
34. Venema, A., van der Ven, L. D. G., and van der Steege, H., *HRC&CC* **2**, 405 (1979).
35. Novotny, M., *J. Chromatogr. Sci.* **8**, 390 (1970).
36. Novotny, M., and Bartle, K. D., *Chromatographia* **7**, 122 (1974).
37. Novotny, M., and Grohmann, K., *J. Chromatogr.* **84**, 167 (1973).
38. Bartle, K. D., and Novotny, M., *J. Chromatogr.* **94**, 35 (1974).
39. van Rijswick, M. H. J., and Tesarik, K., *Chromatographia* **7**, 135 (1974).
40. Grob, K., Grob, G., and Grob, K., Jr., *HRC&CC* **2**, 31 (1979).
41. Cronin, D. A., *J. Chromatogr.* **97**, 263 (1974).
42. Franken, J. J., and de Nijs, R. C. M., *J. Chromatogr.* **144**, 253 (1977).
43. de Nijs, R. C. M., Franken, J. J., Dooper, R. P. M., Rijks, J. A., deRuwe, J. J. M., and Schulting, F. L., *J. Chromatogr.* **167**, 231 (1978).
44. Grob, K., *J. Chromatogr.* **168**, 563 (1979).
45. Gordon, A. L., Taylor, P. J., and Harris, F. W., *J. Chromatogr. Sci.* **14**, 428 (1976).
46. Jennings, W. G., and Wohleb, R., *Chem., Mikrobiol., Technol. Lebensm.* **3**, 33 (1974).
47. Jennings, W. G., Yabumoto, K., and Wohleb, R. H., *J. Chromatogr. Sci.* **12**, 344 (1974).
48. Blumer, M., *Anal. Chem.* **45**, 980 (1973).
49. Schieke, J. D., Comins, N. R., and Pretorius, V., *Chromatographia* **8**, 354 (1975).
50. Sandra, P., Verstaape, M., and Verzele, M., *HRC&CC* **1**, 28 (1978).
51. Donike, M., *Chromatographia* **6**, 190 (1973).
52. Dürbeck, H. W., Büker, I., and Leyman, W., *Chromatographia* **11**, 372 (1978).
53. Schomburg, G., Dielmann, R., Borwitzky, H., and Husmann, H., *J. Chromatogr.* **167**, 337 (1978).
54. Schomburg, G., *HRC&CC* **2**, 461 (1979).
55. de Nijs, R. C. M., Franken, J. J., Dooper, R. P. M., Rijks, J. A., de Ruwe, H. J. J. M., and Schulting, F. L., *J. Chromatogr.* **167**, 231 (1978).
56. Dandeneau, R., and Zerenner, E., *HRC&CC* **2**, 351 (1979).
57. Dandeneau, R., Bente, P., Rooney, T., and Hiskes, R., *Am. Lab.* **11**(9), 61 (1979).
58. Grob, K., *Chromatographia* **7**, 94 (1974).
59. Lee, M. L., Wright, B. W., Phillips, L. V., and Hercules, D. M., Paper No. 4, Expo Chem 79, Houston, Texas, 22–25 October (1979).
60. Schomburg, G., Husmann, H., and Weeke, F., *Chromatographia* **10**, 580 (1977).
61. Grob, K., and Grob, G., *HRC&CC* **1**, 302 (1979).

62. **Janik, A.,** *J. Chromatogr. Sci.* **14,** 589 (1976).
63. **Pierce, H. D., Jr., Unrau, A. M.,** and **Oehlschlager, A. C.,** *J. Chromatogr. Sci.* **17,** 297 (1979).
64. **Papendick, H. D.,** and **Baudisch, J.,** *J. Chromatogr.* **122,** 443 (1976).
65. **Du Plessis, G.,** and **Torline, P.,** *Chromatographia* **10,** 624 (1977).
66. **Desty, D. H.,** and **Douglas, A. A.,** *J. Chromatogr.* **158,** 73 (1978).
67. **Dandeneau, R.,** personal communication (1979).

CHAPTER 3

COLUMN COATING

3.1 General Considerations

Our goal is the deposition of a uniform film, usually 0.1–1.5 μ thick, throughout the column. For very high resolution columns, thinner films are necessary, but their capacities are reduced proportionately (see Section 1.2).

Coating techniques in general can be fitted into one of two general methods, one of which has been termed the dynamic and the other the static technique.

3.2 Dynamic Techniques

The dynamic method, which was first described by Dijkstra and DeGoey [1], as now used by most of its advocates consists of forcing a solution containing approximately 10% liquid phase in a suitable low-boiling solvent through the column under closely controlled flow conditions. Occasionally the column is washed with several column volumes of coating solution, but more generally the coating solution is applied as a single coherent slug, occupying from 2–15 coils of the column, and is forced through the column at a velocity of approximately 1–2 cm/sec with nitrogen pressure. To avoid a higher coating velocity at the outlet end of the column as the back pressure decreases because of the discharge of the slug of liquid phase, a "dummy" column is sometimes attached to the column being coated to serve as a restrictor. Nitrogen flow is continued, and the temperature may be raised gradually to evaporate the remaining solvent. Some workers utilize a single application; others prefer two or three consecutive coating treatments. Blomberg [2] reported that the velocity of the gas during the drying step could exercise an effect on the uniformity of the coating through the column.

Simon and Szepesy [3] in coating their "dehydrated capillaries" (Section 2.2) suggest filling the columns with a 10–30% liquid phase solution and sealing for 24–100 hr. During this period, they reasoned, the larger molecules of liquid phase can diffuse to and adsorb on active sites on the wall. The solution is then removed at a rapid flow rate (0.5–1 m/sec), and the column is dried and conditioned.

Several formulas have been proposed for calculating the final thickness of liquid phase film deposited by the dynamic technique. That given by Novotny and Bartle [4] can be expressed as

$$d_f = \left(\frac{r_0 c}{200}\right) \left(\frac{u\eta}{\gamma}\right)^{1/2} \tag{3.1}$$

where r_0 is the radius of the capillary, c the concentration of liquid phase in the coating solution (vol %), u the velocity of the coating plug, η the viscosity, and γ the surface tension of the coating solution. If during the coating operation the solution is too viscous, the deposited film of liquid phase is too thick and the columns exhibit poor efficiency. Less viscous solutions deposit thinner films, but the lower solution viscosities encourage drainage of the coating solution to the lower surfaces of the column during coating and solvent evaporation. This results in localized thick patches and puddles of liquid phase, which has a drastic and adverse effect on column efficiency.

The interrelationships involved have been considered by Guichon [5]. Parker and Marshall [6] argue that the more serious problem in dynamic coating is the formation of lenses, caused by condensation of solvent vapor in areas behind the coating plug. They attribute this to minor shifts in the column temperature and suggest that the column be subjected to slow temperature programming during coating.

A few liquid phases, including the methyl silicone gum SE 30, have been successfully deposited on untreated glass by these methods, but for more polar liquid phases, a surface pretreatment is usually necessary. The success ratio for the original dynamic methods has been relatively low in the hands of most investigators, probably because of the drainage problems already mentioned. To circumvent this difficulty, Schomburg and Husmann [7] suggested utilizing more concentrated—and hence more viscous—coating solutions, films of which could resist drainage during the drying step. To achieve thin films from these more viscous solutions, they proposed following a short plug of the concentrated liquid phase solution with a few centimeters of mercury, which because of its high surface tension would wipe most of the coating solution off the surface and sweep it out of the column. In practice, these workers utilize a multiphase reservoir (Figure 3.1). The empty etched column, with the coils oriented in the horizontal plane, is attached via a flexible length of heat-shrink

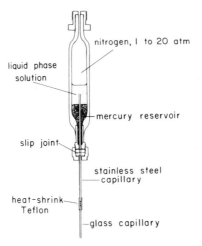

FIGURE 3.1 The Schomburg mercury plug technique [7]. The column, with its coils horizontally oriented, is connected to the stainless steel capillary via a length of heat-shrink Teflon tubing. The stainless steel capillary is lowered so that it feeds from the concentrated liquid phase solution, and the system is pressurized with nitrogen. After a 10–30-cm plug of liquid phase is forced into the column, the stainless steel capillary is adjusted to follow the liquid phase plug with a 3–10-cm mercury plug and then raised to the gas atmosphere. A buffer column is connected to the end of the column being coated to ensure a constant coating velocity.

Teflon tubing to the stainless steel tube. This can be raised or lowered so that it terminates within and feeds from the liquid phase, the mercury, or the gas atmosphere. Nitrogen pressure is applied to the reservoir, and the feed tube is positioned in the liquid phase until some 10 cm of column length are filled with the concentrated solution of liquid phase. The feed tube is lowered to force a small slug of mercury—1–2 cm of column length—into the column immediately behind the slug of liquid phase, and then raised to the pressurized nitrogen atmosphere. To prevent the introduction of a second liquid phase plug behind the mercury plug, the gas pressure should be interrupted while the reservoir feed tube is raised from the mercury to the gas phase. It is critically important that the mercury plug serve as a final wiper. The length of the mercury plug is apparently not critical [8]. After the coating solution and the mercury plug are discharged from the column, the flow of nitrogen is continued while the temperature of the column is increased to remove residual solvent.

3.3   Static Techniques

As currently used by most workers, the static method was developed largely by Verzele and his group [9, 10]. The column is completely filled with a dilute solution (3–10 mg/cm^3) of liquid phase in a suitable low-boiling solvent, and one end is carefully sealed; it is critically important that the column be completely filled with coating solution, and that no air or vapor bubbles exist in the column, especially at the sealed end. The completely filled column is then placed under vacuum, and the solvent evaporated under quiescent conditions leaving a thin film of liquid phase. Obviously, solvent evaporation from either long- or small-bore columns will be a lengthy procedure; typically, the technique is usually restricted to wider-bore (0.5–0.8 mm) columns not exceeding 20–30 m in length. In practice, it is important that the solution be dust-free and degassed to eliminate bumping during the solvent evaporation step. Some recommend that this be accomplished by first filtering a half-strength solution through a microsized membrane filter (*caution:* some filter membranes dissolve in dichloromethane), and then boiling it to half volume to accomplish degassing. The covered solution is cooled rapidly, and the column is filled, preferably by suction, to avoid redissolving gas in the solution.

Others prefer to degas the coating solution by subjecting it to an ultrasonic treatment, usually while applying slight vacuum [11, 12]. The required concentration of the coating solution can be calculated from the relationship

$$d_f = \frac{dc}{400}$$

where d_f is the liquid phase film thickness in μ, d is the column diameter in μ, and c is the volume percent liquid phase [12]. Grob [13] points out that suction filling may give rise to two problems related to evaporation of the low-boiling solvent: The decrease in vacuum may slow the rate of fill, and the increased concentration (hence higher viscosity) at the front end of the solution also slows the fill rate and may even cause a blockage. He suggests the introduction of a short precursor plug of a less volatile solvent immediately preceding the coating solution.

One end of the filled column is lowered so that under the influence of gravity a droplet begins to form, and the completely flooded end is then sealed. Several methods have been suggested to accomplish this, including the attachment of a mechanical plug via heat-shrink Teflon tubing [14]. Others report that a whole variety of sealants can be used with greater reliability [11]. Silicone rubber is preferred by some [15], but authorities such as Grob [13] continue to utilize a water glass seal with excellent results. Sealants such as epoxy resins or other adhesives that may interact with the liquid phase solvent can be shielded with an intermediate aqueous plug. Cueman and Hurley [16] suggest the use of vacuum to pull a 2-cm plug of phenolphthalein water solution into the column directly following the liquid phase solution; a 4–8-cm plug of fast-setting epoxy resin is then drawn in directly behind the water plug and allowed to set.

Precise temperature control is critical during the evaporation step, and most successful practitioners recommend double water baths; i.e., the filled column is immersed in a water bath that is contained in a second water bath that controls the temperature within a very narrow range. Some authorities report that the static coating technique works best when solid or semisolid liquid phases, including the silicone gums, are used with etched glass capillaries. Grob [13] reports that satisfactory columns cannot be prepared from the gum phases such as OV 1, SE 30, SE 52, SE 54, and SP 2125 by dynamic coating and recommends instead static-coating procedures.

Workers that have compared the flow technique with the static method usually feel that better results are achieved with the latter [17, 18], although Novotny concluded that the flow technique was equally satisfactory [19]. Harrison [20] stated that more uniform coatings can be achieved if the column is filled with the calculated amount of stationary phase in benzene, subjected to freezing, and the frozen solvent evaporated by applying vacuum to both ends. This latter method is reported to be less effective with very viscous liquid phases, and according to the authors Carbowax 20 M could not be satisfactorily coated by this technique.

In his original pioneering work, Golay [21] coated columns by completely filling them with a dilute solution of liquid phase in a low-boiling solvent, sealing one end, and drawing the column, open end first, through an oven. As used by Golay, the column could be coiled only after coating, which limited the utility of the method. The method was revised by Ilkova and Mistryukov [22], who conceived the idea of filling the precoiled column with the dilute coating solution and then treating the filled column as if it were a gigantic threaded screw.

The open end of the column is introduced into a high-temperature oven, and the entire column is continuously rotated around its coiling axis; in effect, the column is screwed into the high-temperature oven, forcing solvent evaporation through the open end. When all but the last few coils have been driven into the oven, the sealed end is broken open and the residual solvent vapors removed, either by suction or by dry nitrogen.

One advantage of this modification is that the complete absence of an air space at the sealed end is no longer required and flame-sealing may be utilized. A disadvantage is that a relatively large proportion of the columns prepared by this technique—at least in our hands—is not of the highest quality; examination of the less satisfactory columns shows that thicker annular deposits of liquid phase occur at almost regular intervals (Figure 3.2). When the drying oven is fitted with a glass panel, the cause of this defect is readily apparent: The solvent rarely evaporates smoothly. Several centimeters of column will enter the heated

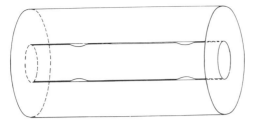

FIGURE 3.2 Thickened annular deposits, occurring at almost regular intervals, can characterize columns dried by the modified Golay technique [22]. This defect is associated with bumping during solvent evaporation and can be corrected with a high-heat-density inlet tube.

zone before the solvent bumps and evaporates in one burst, to be followed by another few centimeters of solution, which evaporate in a similar manner. Obviously, a smoother, more continuous evaporation would be desirable. The realization that this might be better achieved with an abrupt and high-density heat interface led to the development of yet another modification.

The method of Ilkova and Mistryukov (above) was modified by using an arc of stainless steel tubing, heated electrically to a high temperature, as the entrance to the high-temperature oven (Figure 3.3). This much more abrupt heat interface subjects the column

FIGURE 3.3 The heated inlet tube modification of the Golay technique [23, 24]. The abrupt high-density heat interface to which the filled column is subjected on entering the oven creates an entirely different set of coating conditions.

to an entirely different set of conditions during the coating process. When the solution-filled column enters the high-heat-density entrance tube, the solvent is abruptly converted to a superheated vapor containing an aerosol of suspended liquid phase. As the solvent continues to evaporate because of the continual introduction of the solution-filled column into the entrance tube, considerable internal pressure is developed; if a few coils of column are left empty at the closed end of the column before sealing, the compression of the air space at that column end indicates that pressures of 20–30 atm are developed inside the column during coating, although some of that air volume has, of course, disappeared by virtue of its increased solubility in the coating solution as a result of the higher pressure. The interior column wall is subjected to a rather rigorous cleaning treatment and continuously sprayed with the liquid-phase aerosol suspended in high-temperature, high-velocity solvent vapor at high pressure. This may well result in a more intimate contact between the column wall and the liquid phase and may explain the increased stability that has been claimed for these columns [23]. Typically, untreated glass has been used in this technique [24].

Although this last method offers several advantages—it does not require a preetch, any film thickness of any liquid phase can be applied, and columns exhibit superior thermal stability and lower bleed rates—specialized equipment is necessary and any degree of success requires a great deal of experience. Most laboratories that do produce their own glass capillary columns use instead either the dynamic (mercury plug) or static techniques.

Although the cost of the starting materials is much higher, the overall manufacturing costs may be lower for fused silica columns that for columns of conventional glass. The rejection ratio of finished columns is lower, and for a given investment in equipment, labor, and material the manufacturer realizes more saleable columns. Consequently, the price of the fused silica column will probably be lower than prices for conventional glass capillary columns. Even so, they remain labor intensive and must be individually tested. Because of this, the cost of any of the glass capillary columns seems disproportionately high, unless that cost is rationalized on a dollar-per-plate basis, or in terms of a useful lifetime. The true cost of laboratory constructed columns—glass or silica—is invariably much higher and may become prohibitive when the lower efficiencies exhibited by most user-made columns are taken into account. There are enough other specialized problems—sample preparation, injection, detection, and data processing—that add complications to glass capillary GC. The average user interested in the best possible analyses at the lowest possible cost per analysis will find it more economical to direct attention toward these variables and to purchase commercial columns of a guaranteed minimum efficiency.

3.4 PLOT and SCOT Columns

Although a high degree of confusion exists in the use of the terms PLOT and SCOT, a distinction is usually intended. The porous layer of the PLOT column can be achieved with a relatively heavy crystalline deposition or a layer of fused glass powder; the SCOT column also has a porous layer, but it is composed of a support material, usually porous in itself, such as diatomaceous earth. Although all SCOT columns are PLOT columns, not all PLOT columns are SCOT columns.

Although there are exceptions, PLOT columns are frequently constructed in two steps: The porous layer is deposited, and the column is then coated. SCOT columns are more generally made in one step: The coating solution contains the support powder as a suspension. It is not always possible to draw a line of demarcation where surface-pretreated WCOT columns stop and SCOT and PLOT columns begin. Microcrystalline deposits are frequently used with columns that are still described as WCOT columns [25, 26], but as these crystals become larger, some regard them as PLOT or SCOT columns [27]. The particle size used in some PLOT or SCOT columns is even smaller than the crystal deposit occurring in surface-pretreated WCOT columns.

A large number of techniques for preparing these columns have been reported, but coverage here will be restricted to a few representative methods or those that appear to be applicable to glass columns. The pioneering work was done by Halász and Horváth [28], who used coating solutions containing highly dispersed ferric oxide to stabilize a triethylene glycol liquid phase and a finely ground solid support to stabilize squalane in presilvered copper columns. They emphasized that the solvent should have a low boiling point to facilitate its subsequent evaporation and a high density to stabilize the dispersion. Nikelly pulverized Chromosorb W and collected the finer particles by suspension in dry acetone. These were recovered, dried, and 1.5 g suspended in a 2–10% coating solution. The coating solution was placed in a reservoir and forced through a 0.5-mm stainless steel column, which was then conditioned in the usual manner [29]. Blumer [30] used Silanox 101, a fumed silicon dioxide, as the support material. To 10 cm³ of dichloromethane containing 0.2–0.5-g liquid phase was added 0.4–0.65-g Silanox 101, and the suspension was dispersed by brief immersion in a ultrasonic bath. Stainless steel columns were coated by a procedure similar to that described by Nikelly [29]. Kaiser discussed general methods for applying these methods to open tubular glass columns [31]. He pointed out that the support materials covered a very wide range, from naturally occurring or processed biological materials such as starch powder from corn (maize), rice, sporules, or pollen to inorganic powders of aluminum oxide, silicon dioxide, and diatomaceous earths. Because of their large surface areas, such powders may require deactivation, sometimes by pretreatment with a polar liquid phase. Kaiser reported that even highly polar phases such as diethylene glycol succinate (DEGS) can be stabilized on the glass surface with the aid of these fine particles. He utilized a magnetically stirred reservoir to stabilize a suspension of ground Celite and Aerosil in a carbon tetrachloride solution of polypropylene glycol 1500 while it was forced through a horizontally oriented 0.4–0.5-mm × 30-m glass column at a rate of 2 cm/sec.

A method involving precoating the capillary with Silanox 101 followed by a static deposition of liquid phase has been reported to produce columns of increased thermal stability [30, 32]; others have argued that the method increases the sample capacity, but at the cost of resolution.

A novel method for the preparation of PLOT columns, first described by Grant [33], was modified by Cronin [34] to produce what were reported to be highly dependable columns. In Cronin's method, pieces of borosilicate glass with a softening point some 60°C below that of the tubing to be drawn are pulverized and passed through a 200-mesh sieve. One end of a 9-mm o.d. × 3-mm i.d. higher softening point borosilicate glass tube is sealed, and a 2-m length of a carefully cleaned 0.34-mm tungsten wire is placed inside the tube. The glass powder, thoroughly blended 30–70 w/w with 120–150-mesh Celite 545, is poured into the tube around the axially centered wire to achieve a density of 3 g/m. The packed glass tube is then placed in a glass drawing machine with the sealed end extending through the drawing oven. The projecting end of the tungsten wire is then fastened to a fixed anchor so that it can, when fully extended, reach 15–25 mm beyond the oven exit; the free end is then slightly retracted until, still centered in the tube, it is just oustide the entrance to the drawing oven. When the capillary drawing operation is begun, the tungsten wire is picked up and carried along into the oven with the packed glass tube until the anchored end limits further movement. The capillary, with its fused porous layer of powdered glass and diatomaceous earth, is drawn over the tungsten wire and fed through the bending tube to produce a coiled column. The column is then filled with a solution of liquid phase, pressurized at 10–15 atm, drained, dried, and conditioned. A somewhat simpler procedure using "amorphous quartz" particles was described by Torline and Schnautz [35].

3.5 Bonded Phases

Madani *et al.* [36] utilized an entirely different approach. A borosilicate capillary was first etched by the method of Alexander and Rutten [37]. To a chilled flask containing 40 cm³ dichlorodimethylsilane (DCMS), 60 cm³ of 12% ammonium hydroxide was added dropwise with agitation. The polymeric product was recovered some hours later, washed thoroughly with distilled water, and dried by centrifugation. The nearly clear oily fluid was dissolved in dichloromethane (20% v/v) and about one-fourth of the column filled with the solution. Nitrogen pressure was applied to remove the excess solution, and gas flow continued for several hours to dry the deposit. The column was then connected to a flask containing solid sodium hydroxide and 25% ammonium hydroxide added to force gaseous ammonia through the column. After sealing both ends, the column was heated to 320°C for 24 hr, flushed clean, and conditioned. The columns reportedly achieved excellent separations on a number of test mixtures. More recently, Madani and Chambaz reported that a reactive methyl-phenyl-polysiloxane polymer could be chemically bonded to the surface of a glass capillary by applying a similar technique [38].

3.6   Packed Capillary and Micropacked Columns

Common usage defines the packed capillary column as one produced by drawing out glass tubes that have first been loosely filled with a granular support material of narrow mesh range [39], while micropacked columns are columns of similar dimensions that are filled with a precoated packing after drawing [40]. The former are restricted to packings of high mechanical strength and great thermal stability; no such restrictions are imposed on the latter. Both types of column can be considered as intermediates whose capacities and efficiencies lie between the extremes exhibited by packed and WCOT columns. In general, the internal diameter is less than 1.0 mm, and a number of advantages have been claimed: Pressure drop through the column is relatively low, the separation efficiency is relatively high, and direct injection (*vide infra*) poses no difficulty. With micropacked columns, any support material and any liquid phase can be used. The increased capacity of these columns makes them quite attractive for direct coupling to mass spectrometers or for other application in which it is desirable to recover isolated fractions or subject the effluent to sensory analysis.

Kaiser [41] discussed the advantages offered by these columns, and Bruner *et al.* [42] described a method of packing with graphitized carbon black. In a later study [43] Bruner *et al.* used this method to prepare micropacked columns of *n* values up to 30,000 and *h* equal to 0.45 mm on which the separation of a variety of products was demonstrated. Berezkin *et al.* [44] prepared short lengths of micropacked columns and reported the effect of column diameter and carrier gas (helium and nitrogen) on their performance characteristics. Guichon [45] and Rijks [46] discussed both the preparation and use of micropacked columns, and their increased sample capacity has been thoroughly explored [47].

Regardless of the type of column under preparation, coating solutions should in most cases be freshly prepared. Columns prepared from stock solutions only a week or two old sometimes exhibit anomalous behavior [48]; this phenomenon may relate to the fact that degradation of several silicone liquid phases has been demonstrated in stock solutions that used chlorinated hydrocarbons as the solvent [49, 50].

References

1. Dijkstra, G., and DeGoey, J., *in* "Gas Chromatography 1958" (D. H. Desty, ed.), pp. 56–58. Butterworth, London, 1958.
2. Blomberg, L., *Chromatographia* **8**, 324 (1975).
3. Simon, J., and Szepesy, L., *J. Chromatogr.* **119**, 495 (1975).
4. Novotny, M., and Bartle, K. D., *J. Chromatogr.* **93**, 405 (1974).
5. Guichon, G., *J. Chromatogr. Sci.* **9**, 512 (1971).
6. Parker, D. A., and Marshall, J. L., *Chromatographia* **11**, 526 (1978).
7. Schomburg, G., and Hussmann, H., *Chromatographia* **8**, 517 (1975).
8. Alexander, G., and Lipsky, S. R., *Chromatographia* **10**, 487 (1977).
9. Bouche, J., and Verzele, M., *J. Gas Chromatogr.* **6**, 501 (1968).
10. Verzele, M., Verstappe, M., Sandra, P., van Lüchene, E., and Vuye, A., *J. Chromatogr. Sci.* **10**, 668 (1972).
11. Giabbai, M., Shoults, M., and Bertsch, W., *HRC&CC* **1**, 277 (1978).
12. Rutten, G. A. F. M., and Rijks, J. A. *HRC&CC* **1**, 279 (1978).
13. Grob, K., *HRC&CC* **1**, 93 (1978).
14. Sandra, P., and Verzele, M., *Chromatographia* **11**, 102 (1978).
15. Roeraade, J., personal communication (1977).
16. Cueman, M. K., and Hurley, R. P., Jr., *HRC&CC* **1**, 92 (1978).
17. Roeraade, J., *Chromatographia* **8**, 511 (1975).
18. Sandra, P., Verzele, M., and van Lüchene, E., *Chromatographia* **8**, 499 (1975).
19. Novotny, M., *J. Chromatogr. Sci.* **8**, 390 (1970).
20. Harrison, I. T., *Anal. Chem.* **47**, 1211 (1975).
21. Golay, M. J. E., *in* "Gas Chromatography 1958" (D. H. Desty, ed.), pp. 36–55. Butterworth, London, 1958.
22. Ikova, E. L., and Mistryukov, E. A., *J. Chromatogr. Sci.* **9**, 569 (1971).
23. Jennings, W. G., and Wohleb, R., *Chem., Mikrobiol., Technol. Lebensm.* **3**, 33 (1974).
24. Jennings, W. G., Yabamoto, K., and Wohleb, R. H., *J. Chromatogr. Sci.* **12**, 344 (1974).
25. Schulte, E., *Chromatographia* **9**, 315 (1976).
26. Nota, G., Goretti, G. C., Armenante, M., and Marino, G., *J. Chromatogr.* **95**, 229 (1974).
27. Watanabe, C., and Tomita, H. J., *Chromatogr. Sci.* **13**, 123 (1975).
28. Halász, I., and Horváth, C., *Anal. Chem.* **35**, 499 (1963).
29. Nikelly, J. G., *Anal. Chem.* **44**, 633 (1972).
30. Blumer, M., *Anal. Chem.* **45**, 980 (1973).
31. Kaiser, R., *Chromatographia* **1**, 1 (1968).
32. van Hout, P., Szafranek, J., Pfaffenberger, C. D., and Horning, E. C., *J. Chromatogr.* **99**, 103 (1974).
33. Grant, D. W., *J. Gas Chromatogr.* **6**, 18 (1969).

34. **Cronin, D. A.,** *J. Chromatogr.* **48**, 406 (1970).
35. **Torline, P.,** and **Schnautz, N.,** *HRC&CC* **1**, 301 (1978).
36. **Madani, C., Chambaz, E. M., Rigaud, M., Durand, J.,** and **Chebroux, P.,** *J. Chromatogr.* **126**, 161 (1976).
37. **Alexander, G.,** and **Rutten, G. A. F. M.,** *Chromatographia* **6**, 231 (1973).
38. **Madani, C.,** and **Chambaz, E. M.,** *Chromatographia* **11**, 725 (1978).
39. **Halász, I.,** and **Heine, E.,** *Anal. Chem.* **37**, 495 (1965).
40. **Cramers, C. A., Rijks, J. A.,** and **Bocek, P.,** *J. Chromatogr.* **65**, 29 (1972).
41. **Kaiser, R.,** *J. Chromatogr.* **112**, 455 (1975).
42. **Bruner, F., Ciccioli, P.,** and **Bertoni, G.,** *J. Chromatogr.* **90**, 239 (1974).
43. **Bruner, F., Ciccioli, P., Bertoni, G.,** and **Liberti, A.,** *J. Chromatogr. Sci.* **12**, 758 (1974).
44. **Berezkin, V. G., Shkolina, L. A.,** and **Svyatoshenko, A. T.,** *J. Chromatogr.* **99**, 111 (1974).
45. **Guichon, G.,** *Adv. Chromatogr.* **8**, 179 (1969).
46. **Rijks, J. A.,** Doctoral Thesis, Tech. Univ., Eindhoven, Netherlands, 1973.
47. **Pretorius, V.,** and **Smutts, T. W.,** *HRC&CC* **2**, 444 (1979).
48. **Grob, K.,** and **Grob, G.,** *HRC&CC* **1**, 221 (1978).
49. **Venema, A., van der Ven, L. G. J.,** and **van der Steege, H.,** *HRC&CC* **2**, 69 (1979).
50. **Venema, A., van der Ven, L. G. J.,** and **van der Steege,** *HRC&CC* **2**, 405 (1979).

CHAPTER 4

INLET SYSTEMS

4.1 General Considerations

No degree of column excellence can compensate for design defects in the chromatographic system. Minor design defects that may not have been apparent with the higher carrier gas flows used with packed columns may become intolerable when a column of capillary dimensions is installed. Particular attention should be directed to areas of excessive volume and dead space that most usually occur where the column is attached to the inlet and the detector.

Columns capable of tolerating carrier gas flow rates of 5–8 cm^3/min—including micropacked and most PLOT, SCOT, and larger-bore WCOT columns—can usually utilize normal gas chromatographic inlet systems, although models with glass vaporization surfaces are to be preferred because of their lower reactivity. Small-bore capillaries, however, achieve their highest efficiencies at relatively low flow rates. Columns of 0.25-mm i.d. may have an optimum average carrier gas velocity as low as 6–12 cm/sec with nitrogen carrier gas (a poor choice under most circumstances; see Chapter 8), equivalent to a flow rate of 0.18–0.35 cm^3/min. Even if the injection or vaporization chamber has a minimum volume—perhaps 1 cm^3—and assuming that a single volume of gas is sufficient to flush it clean—which is highly doubtful—it would require from 3 to 5 min to sweep the volume of the vaporization chamber to the column. The band of material deposited on the beginning of the column would be 3–5 min wide, and each substance that begins the chromatographic process immediately on injection will produce a peak at least 3–5 min wide, an occurrence that has caused more than one uninformed initiate to reject WCOT columns and return to the world of packed columns. Even if helium were used as the carrier gas and supplied at the optimum practical gas velocity (Chapter 8), almost a full minute would be required to sweep the sample vaporization chamber with a single volume of gas. This would lead to an unacceptable band width at the beginning of the process; several injection techniques are used to overcome this problem.

4.2   Cold Trapping (Precolumn Concentration)

If the column is maintained at a lower temperature so that the distribution constants of all injected components are very large, their concentrations in the gas phase are negligibly small. These conditions, in which the injected components concentrate in the liquid phase at the beginning of the column and fail to chromatograph, must be maintained until the vaporization chamber has been flushed clean. The column temperature may then be increased, and the tight band at the beginning of the column will then begin the chromatographic process. Aspects of this method are considered further in Chapter 12.

In general, substances whose boiling points are at least 50°C (and preferably more than 100°) higher than the column temperature can be effectively concentrated on column by cold trapping techniques. For special applications only a short section of the column may be cooled to the required temperature; with the glass capillary column this is conveniently accomplished by bending a U-tube in the column as it leaves the inlet (Figure 4.1). This is located so that with the oven door open, a beaker of coolant (selected with attention to what materials are to be trapped) can be inserted, raised, and blocked so that the U-tube is immersed in the coolant during the trapping period. The coolant is then removed, the oven door closed, and the chromatographic process

FIGURE 4.1 A glass capillary with U-tube for specialized cold-trapping application.

begun. For higher-boiling compounds cold trapping can be carried out in the column without any special adaptation [1, 2].

Trap-and-purge methods of sample preparation usually depend, in the final analysis, on cold trapping to reconcentrate on column the collected sample as it is purged from the trapping substrate (Section 12.10).

In those injection techniques that involve on-column concentration of carrier-gas-borne materials for some period of time (i.e., cold trapping, "splitless," and some forms of on-column injection), residual high-boiling substances in the vaporization chamber and compounds bleeding from the septum will be trapped with the sample volatiles. A septum-bleed (or septum-purge) cap can do much to alleviate this problem (Figure 4.2).

It may be worth noting that this same injection technique is a frequent inadvertent occurrence and is often the cause of the

FIGURE 4.2   A commercial septum-bleed cap that replaces the normal septum-retaining cap. A self-contained needle valve permits adjustment of the purge gas flow.

"ghost peaks" that occur in many cases of temperature programming. Higher-boiling contaminants from the carrier gas, flow controllers, connecting tubing, or septum bleed slowly concentrate on the front end of the cold column and begin the chromatographic process as the column temperature is increased, giving rise to what have been termed ghost peaks, or "Lorelei peaks" (*ich weiss nicht was soll es bedeuten*). The various causes of these extraneous peaks and methods for their reduction or elimination were discussed by Purcell *et al.* [3]. Suggested solutions included a septum cap with a sliding seal so that the silicone rubber was in contact with the carrier gas only at the time of injection, careful cleaning of connecting tubing and flow controllers, and the installation of a trap in the carrier gas line just preceding the inlet.

4.3   Splitless Injection

Some have suggested that the lower-temperature concentration technique just considered is the underlying principle of "splitless injection" [4], except that an extra provision, in the form of a backflush or purge stream, is usually added to flush the last injection residues from the inlet before the chromatographic process is begun. Splitless injection, however, utilizes a "solvent effect" that was first analyzed by Harris [5] and developed to a very

high degree by Grob. Although the concept of precolumn concentration combined with the solvent effect (*vide infra*) has been
used by several investigators, true splitless injection can be
achieved isothermally and, as Grob and Grob have emphasized
[6, 7], is not analogous to cold trapping. The solvent should be
selected with attention to its boiling point as related to the initial
column temperature, the type of liquid phase, and the nature of
the sample components. It is essential that the solvent condense
on the first section of the column, creating a serious overload
condition that permits it to play the role, temporarily, of liquid
phase [8]. If the boiling point of the solvent is too low with respect
to column temperature, this condition cannot be achieved. On
the other hand, if the boiling point of the solvent is too high, its
subsequent separation from the sample components during the
chromatographic process may not be realized (Table 4.1).

Splitless injection, then, results in a tremendous (but tempo-
rary) decrease in the β value of the column, if we regard the
condensed solvent as liquid phase. If the solvents have affinities
for the selected solvent that are not widely different from their
affinities for the stationary phase, their individual K_D values
remain reasonably constant. Because the individual K_D values of
the sample components remain "constant" and the β of the column is very much lower, all k values are greatly increased (Figure
4.3). Consequently, each sample component is much more

TABLE 4.1

**Recommended Initial Temperatures for Splitless
Injection[a]**

Solvent	Boiling point (°C)	Recommended initial temp (°C)
Dichloromethane	40	10–30
Chloroform	61	25–50
Carbon disulfide	46	10–35
Diethyl ether	35	10–25
Pentane	36	10–25
Hexane	69	40–60
iso-Octane	99	70–90

[a] From T. A. Rooney, "Optimizing Analyses Using Splitless Injection on Capillary Columns," Application Note
AN 228-5. Hewlett Packard Co., Avondale, Pennsylvania.

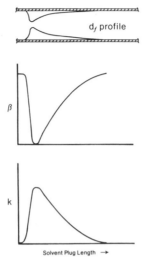

FIGURE 4.3 Relationship of the solvent effect to fundamental parameters influencing the partitioning process in the column; $\beta = r_0/2d_f$, $K = K_D/\beta$. The plug of condensed solvent assumes a configuration as shown at the top [8]; this changes the phase ratio of the column (center), which forces a shift in the partition ratios (bottom). (From Jennings *et al.* [9].)

strongly associated with the liquid phase, and the number of solute molecules in the moving gas phase is greatly reduced [Eqs. (1.1), (1.7), (1.9), and (1.10)]. The front of the sample plug moving forward through the column continues to encounter an ever increasing concentration of liquid phase, and as a consequence is more strongly retained than is the rear of that plug; hence band narrowing should occur during properly executed splitless injection [9]. As the solvent plug moves out, which is usually encouraged by increasing the column temperature, the normal β value is recovered, the sample components reassume their normal k values, and the chromatographic process proceeds.

The condensed solvent should take the form of a zone of high solvent concentration. Formation of a true solvent plug that actually bridges the column can lead to rapid column deterioration. Peaks preceding the solvent peak (or indeed, any other major peak chromatographed under conditions where it can affect the phase ratio of the column) experience a reverse solvent effect [10];

in extreme cases this can lead to complete degradation of peak shape (Figure 4.4).

To minimize the column deterioration that can result from solvent overloading, a solvent in which the liquid phase is not readily soluble should be selected. Methanol appears to be a particularly poor choice for a solvent in splitless injection, in that

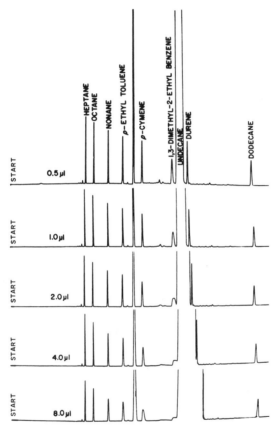

FIGURE 4.4 Normal and reverse solvent effects, here illustrated in split mode, are experienced to an even greater degree in "splitless" injections. Indicated injection volumes were each split 1:100 and contained 0.08 vol % of each labeled test substance in undecane. Column 0.25-mm i.d. × 50 m, OV 101, d_f = 200 nm; isothermal at 125°C, helium carrier at 20.5 cm/sec. Associated data are shown in Table 4.2. (From Miller and Jennings [10].)

on-column condensation of methanol can lead to rapid column deterioration. Some users report that when glass capillaries coated with the methyl silicone SP 2100 are repeatedly subjected to injections involving the condensation of methanolic solutions, the columns lose efficiency fairly rapidly to a point where they possess ~50% of their original theoretical plates; efficiency continues to decline, but at a much slower rate as the methanol injections are continued. Columns coated with SP 1000 are considerably more resistant to methanol, and even under repeated splitless injections of methanolic solutions their loss in efficiency is limited to ~10%. Columns coated with the silicone gums SE 30, SE 52, or SE 54 seem somewhat more tolerant of methanol condensation, according to user reports. This suggests that where the superior coating efficiency of the SP 2100 is necessary and splitless injections of methanolic solutions are required, one might consider attaching a short SE 30 or SE 54 precolumn to the SP 2100 analytical column, preferably using a short platinum–iridium junction (Chapter 8). Use conditions should be such that the solvent plug condenses in the precolumn and the methanol, as it reaches the analytical column is sufficiently dispersed that it can no longer bridge the column.

Acetone is another solvent to be avoided with silicone liquid phases, as it can lead to rapid deterioration of methyl silicone columns. Hydrocarbons and dichloromethane have both been widely used as solvents in splitless injection, and good results have been reported with carbon disulfide, although the author has some intuitive reservations about the suitability of the latter because of its strong adsorptive—and hence displacement—tendencies.

Because a finite period is allowed for the injected sample to be swept to the column, vaporization of the sample can proceed more slowly and lower inlet temperatures are possible with splitless injection. The volume of the vaporization chamber should be small enough that it can be swept in a reasonable period of time, but it must be large enough to accommodate the vaporized sample or flashback of the latter can contaminate the septum, carrier gas line, and so on. A volume of ~1 cm^3 is widely used, usually involving a glass insert designed to minimize unswept passages. The large volume of solvent injected can lead to severe solvent tailing, which obliterates most of the chromatogram. Recognizing that only a small amount—perhaps 5-10%—of the injected solvent accounts for the massive solvent tail, Schulte and Acker [11] suggested a method for purging from the system this last portion of the injection (Figure 4.5). Various modifications of this approach are used in most modern splitless injectors. If the purge system is activated too soon, a major portion of the injected sample is lost; if it is activated too late, excessive tailing occurs. Most systems yield satisfactory results if the purge function is activated after a volume of carrier gas equal to ~1.5 times the volume of the vaporization chamber has swept through the latter (Figure 4.6). With the beginning of solvent condensation in the column, the relatively large volume of solvent vapor in this area condenses to a smaller volume of liquid, leading to a rapid and localized drop in pressure that actually helps draw material out of the vaporization chamber.

It should be emphasized that because of this necessary purge function, "splitless" injection is not truly splitless, and a portion of the injection is split under conditions that make it difficult to check the linearity of that split.

The critical relationship between the initial temperature and the boiling point of the solvent has already been discussed. In addition, the volume of solvent injected must be large enough (1–3 μl) to achieve the solvent effect. Blank runs should always be

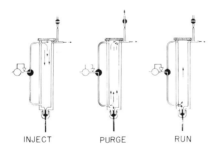

INJECT   PURGE   RUN

FIGURE 4.5 The Schulte and Acker inlet, shown in splitless mode. The injection is made with the purge function (top valve) off, and carrier gas sweeping the full length of the injection chamber, thence to column. Purge function is accomplished by redirecting the carrier flow to enter the bottom of the injection chamber, whence a minor portion enters the column and a major portion purges the injection chamber via the opened purge valve (top). The purge valve is then closed; septum sweep is operative at all times. (Adapted from Schulte and Acker [11].)

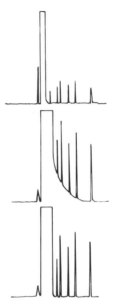

FIGURE 4.6   Influence of purge activation time. Top, purge activated too soon, a restricted amount of sample (and solvent) has been transported from the inlet to the column. Center, purge delayed too long, and last portion of sample (and solvent) was carried to column (heavily diluted with carrier gas) after plug began to move. Bottom, proper purge activation time.

used to evaluate solvent purity. Because splitless injection allows a longer time for vaporization of the injected sample and its transfer to the column, lower inlet temperatures can be utilized (typically 200–250°C), discouraging both sample degradation and septum bleed. Curious effects are sometimes noted when splitless injections are used with hydrogen carrier gas at reasonably high velocities (50–60 cm/sec); solute peaks can exhibit doublets or even triplets under these conditions. With the substitution of helium or nitrogen under these same temperature and velocity conditions, this effect disappears [12]. The cause of this anomalous behavior has not yet been defined but almost surely relates to the carrier gas densities and pressures. Splitless injection finds its greatest utility where component concentrations are too low for detection by split injection (i.e., less than 0.1% of the sample) or where only a very limited amount of sample is available.

4.4   Inlet Splitters

One of the simplest and most popular methods used to avoid the deposition of broad bands on the column utilizes an inlet splitter. Splitting the flow stream permits one to use a high flow rate through the vaporization chamber, so it is rapidly flushed clean while still maintaining the restricted flow volume through the column. Contrary to popular opinion, the major function of the inlet splitter is not only to restrict the size of the sample placed on the column; its frequently more important function is to permit the rapid flushing of the injection chamber so the sample on the column is followed by pure carrier gas rather than by exponentially diluted sample. A number of splitter designs have been described [13, 14], and criteria for the evaluation of splitter linearity—which have been frequently abused in commercial comparisons—are also available [15, 16]. Linearity refers to the degree of fidelity achieved as the splitter directs only a portion of the injected sample to the column. A splitter with poor linearity demonstrates discrimination and may emphasize low-boiling components at the expense of higher-boiling constituents, or vice versa. Much of the data used to advertise various commercial inlet splitters demonstrates reproducibility but not linearity.

Splitter linearity can be checked in a very unambiguous manner. The splitter is left in position but the capillary column is removed and a packed column substituted. The splitter outlet is turned off or capped so that the entire flow is directed through the column, and the make-up gas is shut off at the detector. Under these conditions where the entire injection is delivered to the column and thence to the detector, the detector is carefully optimized and a model system is prepared on which this packed column is able to achieve baseline resolution and that represents the range of functional groups and boiling point extremes of samples later to be analyzed on the capillary column. The mixture should be injected several times, to satisfy any demand capacity of the system, and when reproducible results are obtained, the normalized peak areas from several runs should be averaged. The packed column is then removed, the capillary reinstalled, and the split ratio adjusted (*vide infra*). Make-up gas is supplied to the detector to compensate for the difference between the packed and capillary column flows, and the model system is injected. By comparing these results with the former, one can draw some meaningful conclusions relative to splitter linearity.

Where linearity is of prime importance, split ratios lower than 1:50 (with helium carrier gas) may pose a problem. It may be wise to use a split ratio of at least 1:100 to establish the fidelity of that splitter and then to change the split ratio to see how far this can be lowered before linearity falls off. Watanabe *et al.* [17] analyzed in some detail the linearity of selected splitter designs.

Both normal and reverse solvent effects can occur in split injection, even with small injections and high split ratios [8–10]. Figure 4.4 shows chromatograms representing different size injections, all at split ratios of 1:100, and Table 4.2 shows plate numbers calculated on the various peaks. Those peaks exhibiting an increase in plate number (i.e., a narrower peak) as a function of injection size have experienced the normal solvent effect, while those demonstrating a decrease in plate number (i.e., peak broadening) as a function of injection size have experienced a reverse solvent effect. In extreme cases the latter has resulted in complete degradation of the peak shape. Several authors [18, 19] have discussed design characteristics for inlet splitters. Until quite recently most splitters capable of good linearity possessed metallic sample-contact surfaces, although it was recognized that rearrangement and/or degradation of some sample constituents occurred. This, of course, would sacrifice a major advantage of the glass capillary column: inertness. Figure 4.7 illustrates a simple concentric tube type of inlet splitter easily constructed of glass, and this same design is used in the assembly shown in Figure 4.10. Unfortunately, severe discrimination can be demonstrated on difficult samples with this simple type of splitter. Ideally, the sample should be vaporized, subjected to a high degree of mixing, and then undergo expansion just prior to splitting. Sufficient buffer volume should be provided beyond the split point to minimize pressure fluctuations that would otherwise result from sample vaporization. It is also desirable that the injected sample contact only glass surfaces until after the splitting occurs, and that the glass chamber(s) in which vaporization, mixing, and expansion occur be removable for periodic cleaning and inactivation (e.g., silanization). Figure 4.8 shows an inlet splitter that incorporates most of these features [20], and Figure 4.9 shows a highly linear all-glass inlet system [19, 21].

TABLE 4.2

Theoretical Plate Numbers Exhibited by Test Substances at Different Sample Injection Volumes[a]

Solute	k[b]	0.015 μl	0.03 μl	0.06 μl	0.12 μl	0.25 μl	0.50 μl	1.0 μl	2.0 μl	4.0 μl	8.0 μl	Pentane test[b] solution
n-heptane	0.166	4400	4400	4400	4400	4400	4400	4400	4400	3700	2700	4400
n-octane	0.295	11640	11640	11640	11640	11640	11640	11640	11640	8550	5200	11640
h-nonane	0.521	25500	25500	25500	25500	25500	25500	25500	25500	13000	6650	25500
p-ethyl toluene	0.742	36500	36500	36500	36500	36500	36500	36500	36500	16900	8140	36500
p-cymene	1.037	51600	51600	51600	51600	51600	51600	51600	30525	1170	NM	51600
1,3-DM-2-EB	1.500	74900	74900	63800	59100	42100	31500	19550	NM[c]	NM	NM	74900
durene	1.742	86000	86000	86000	93100	110150	120200	132000	180000	350000	720000	86000
n-dodecane	2.790	120400	120400	120400	120400	120400	120400	120400	120400	120400	120400	120400

[a] (See Figure 4.4.) From Miller and Jennings [10].
[b] Partition ratios shown in column 2, and effective theoretical plate numbers shown in column 13 were determined on a pentane solution containing 0.08 vol % of each test substance; neither the normal or reverse solvent effect should occur with this solution.
[c] NM = not measurable.

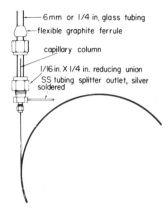

FIGURE 4.7   Simple concentric tube splitter. Normally, the upper end of the 6-mm (or ¼-in.) glass tube is packed with fine glass beads (silanized), the flexible graphite ferrule is slipped downward, and as much of the splitter tube as possible—including that portion where splitting occurs—is housed within the heated chamber of the normal "on-column injector" of the gas chromatograph. The $\frac{1}{16}$-in. splitter outlet tubing may be formed into a column hanger before proceeding to the restrictor. The restrictor may take the form of a needle valve, a short dummy column, or a constriction in the tubing.

At high split ratios, the efficiency of sample vaporization can be affected adversely; the high carrier gas flow rate will tend to cool the vaporization surfaces unless the carrier gas is subjected to adequate preheating. In the absence of complete vaporization, components of the injected sample are present as a fog or aerosol,

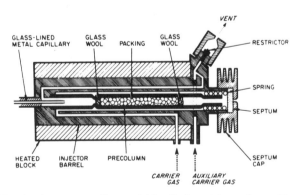

FIGURE 4.8   A commercially available all-glass inlet splitter. (See Ettre [18, 20].)

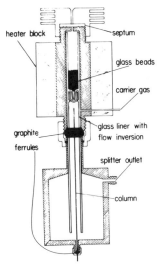

FIGURE 4.9 A highly linear all-glass splitter. Originally described by Jennings [19], and later modified to adapt a conventional heated ¼-in. on-column injector [21].

and the composition at the split orifice is neither constant through the injection nor representative. This of course has a disastrous effect on linearity.

Some inlets include a provision for deactivating the splitter during the run, permitting the conservation of carrier gas. In this case it is advisable to measure the carrier gas velocity through the column under those conditions that prevail during the run, (i.e., with the splitter inactive). When the closed splitter outlet is first opened, pressure in the inlet—and on the front of the column— drops. Carrier gas in the front portion (i.e., at the inlet end) of the column will flow in the reverse direction while the column adjusts to this new and lower pressure drop. (This phenomenon may also cause problems when the purge valve is opened, with improperly designed equipment, during splitless injection as already described.) Therefore it is desirable to activate the splitter 1–2 min before injecting the sample; additionally, the splitter outlets should be left open for precisely the same period of time— which usually is within the range 15 sec to 2 min—each time after injecting. Even when the splitter is inactive, a very low bleed (1– 2 cm³/min) should be discharged from the splitter outlet. This discourages back diffusion of the discarded split stream, a condition that can lead to extraneous peaks and severe noise problems. Figure 4.10 shows a diagram of such a splitter.

Most of the splitters illustrated to this point (i.e., Figures 4.7–4.10) require straightening the end of the capillary so that it can be inserted into the splitter. For a variety of reasons the use of unstraightened columns appeals to many workers, and several devices have been suggested for accepting unstraightened columns. In several cases these contain a design defect in the form of a butt joint in a low-velocity gas zone (Figure 4.11). Figure 4.12 shows an alternative approach, in which an unstraightened column terminates in a high-velocity gas zone. The disadvantage in this latter design is that the split point is at oven temperature, and discrimination is evidenced in some of the high-boiling components (e.g., boiling points above 350°C) of samples such as crude oils; these problems are less severe at higher split ratios.

Figure 4.13 illustrates an all-glass splitter, accepting unstraightened columns, that offers some advantages to those engaged in the analysis of flavors, essential oils, and some other samples. Many materials, including most natural products, contain nonvolatile components that gradually accumulate on the vaporization

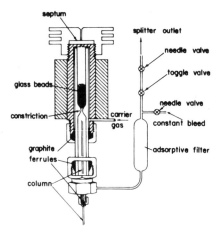

FIGURE 4.10 Diagram of a splitter capable of intermittent operation. It is advisable to install an adsorptive filler as shown in every case in which an inlet splitter is used.

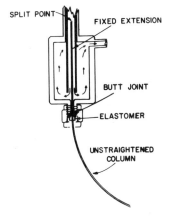

FIGURE 4.11   Schematic of a splitter designed for use with unstraightened columns, with fixed extension and butt joint in a low-velocity gas zone.

surfaces. These deposits slowly oxidize, become acidic, and can react with subsequent injections to institute a variety of isomerization and degradation reactions. With this all-glass splitter, any deposits are readily apparent; the device is exchanged and the old unit cleaned by soaking in cleaning solution.

For most purposes splitters can be deactivated by silanization, and this is usually best accomplished *in situ*. The splitter outlet

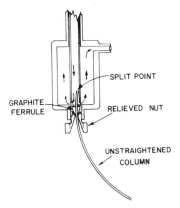

FIGURE 4.12   Schematic of a splitter designed for use with unstraightened columns, with column end in high-velocity gas zone. The end of the column should not be in contact with the glass liner.

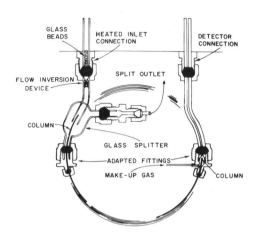

FIGURE 4.13   Schematic of an all-glass splitter, designed for use with un-straightened columns. (Courtesy T. Shibamoto, Ogawa and Co., Ltd.)

should be turned off or capped and the column removed and the column fitting capped. The inlet and oven temperatures are set at 150°–200°C, and ~8 μl of a 50–50 mixture of dichlorodimethylsilane and hexamethyldisilazane is injected three or four times at 15-min intervals. About 60 min after the final injection, one end of a tight-fitting tube is connected to the splitter outlet while the other end is immersed in a beaker of water; the splitter outlet is opened and the system is flushed with dry carrier gas. Alternatively, a wet cloth can be packed around the splitter outlet while the system is flushed. Silanization is not an irreversible process, and depending on the dryness of the carrier gas and the composition and susceptibility of the sample, the process may have to be repeated periodically.

Some workers prefer to deactivate splitter components by immersion in dichloromethane containing 10 mg/cm³ Carbowax 20 M. The glass components are then drained and allowed to air-dry, and the splitter is installed. Again the process may have to be repeated after a period of use.

The splitting behavior of an inlet splitter can be influenced by the nature of the sample (i.e., the nature of the solutes and their range of boiling points), the size of the injection, the temperature of the inlet, the nature of the carrier gas, and the split ratio. The temperature of the splitter should if possible approach the boiling point of the higher-boiling components in the system. The anomalous behavior that can otherwise occur is most pronounced at lower split ratios. Figure 4.14 illustrates this point with a C_{13}–C_{16} paraffin hydrocarbon mixture, whose boiling points range from 235° through 287°C; the inlet, of course, is under pressure and the boiling points under these conditions would be appreciably higher. The inlet in this case was operated at 250°C. At the lower split ratios a marked departure from linearity is apparent, which takes the form of discrimination against the lower-boiling and in favor of the higher-boiling components. The latter vaporize more slowly and at this low split ratio are deposited on column for a longer period of time.

At split ratios exceeding 1:100, discrimination is much less, even with the inlet at this low temperature. As the temperature of the inlet is increased, the influence of the split ratio becomes less pronounced. At 275°C linearity is improved at all split ratios, and at 300°C there is essentially no difference between split ratios of 1:100 and 1:400. Even at 300°C departures from linearity can be detected at very low (1:10) split ratios. This type of discrimination can be quite consistent, emphasizing that mere replication of results is not a good criterion of splitter linearity.

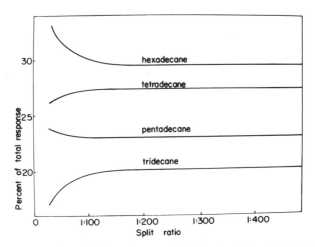

FIGURE 4.14 Effect of split ratio on splitter discrimination. Splitter temperature, 250°C. Test mixture, tridecane (b.p. 235°C), tetradecane (b.p. 254°C), pentadecane (b.p. 271°C), and hexadecane (b.p. 287°C).

Ignoring sample composition, then, the two factors that exercise the greatest effect on splitter linearity are the split ratio (or flow velocity through the splitter) and the temperature. At a low flow rate, splitters with smaller internal passages and volumes should achieve a higher degree of turbulence before the vaporized sample is subjected to expansion and splitting; hence low split ratios would be less deleterious to their performance. If only a restricted amount of sample is available and a high split ratio cannot be tolerated (*vide infra*), the temperature of the inlet may be increased to improve linearity. With samples that exhibit a high degree of thermal lability, lower inlet temperatures can be compensated for by higher split ratios. On-column injection may offer the best solution where a limited amount of sample containing an unstable compound is to be analyzed.

Because retention measurements can yield much more information on these high-resolution columns, fluctuations in carrier gas flow (which cause fluctuations in retention time behavior) can be especially exasperating. In most installations the capillary column is operated at constant pressure drop; flow through the column is a function of the column head pressure and the column temperature. Hence even though the flow rate through the column varies with temperature programming, flow rates, retentions, and relative retentions are comparable from run to run as long as the pressure drop, temperature and program rate, phase ratio (amount of liquid phase), and length of column remain constant (Section 8.8). Condensation in the needle valve (Figure 4.15) or non-steady-state leaks at the septum are those problems that most

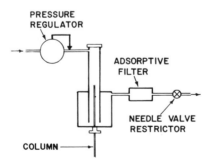

FIGURE 4.15 Schematic of carrier gas flow through splitter utilizing pressure-regulated inlet and restrictor outlet.

frequently cause fluctuations in column flow that lead to retention shifts in the capillary system. Figure 4.16 shows a method of dealing with this. There are normally two outlets from the pressurized inlet, one through the column and one through the splitter outlet. The flow rate through the column depends on the pressure in the inlet and on the relative restrictions offered by these two outlets. The split ratio is varied by changing the degree of restriction offered by the splitter outlet, which also changes the column head pressure. A leaky septum would offer a third and variable outlet, which would also change the column head pressure if the gas was supplied by pressure regulation and discharged from the splitter through a restriction (Figure 4.15).

By using a flow controller to supply carrier gas to the inlet and discharging the splitter outlet through a back-pressure regulator, column head pressure can be held constant even under conditions of a leaky septum, provided that gas continues to discharge through the back-pressure regulator (Figure 4.16). The flow controller governs the total flow through the inlet (and determines the split ratio), and the back-pressure regulator controls the flow through the column. A septum leak, for example, will in this case cause a change in the split ratio, but not in the column flow. To utilize this system properly, the adsorptive filter in the discharge line should have an extremely small pressure drop; a short, wide, and loosely packed filter is, for these purposes, superior.

Inlet splitters suffer some prejudice from the fact that the major portion of the injected sample is discarded. The individual studying a trace component whose accumulation has required considerable expense and effort may be loath to discharge to the atmos-

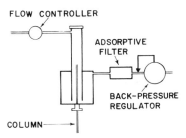

FIGURE 4.16 Schematic of carrier gas flow through splitter utilizing flow-controlled inlet and back-pressure-regulated outlet.

phere the greatest part of the injected sample. The use of an inlet splitter with a fine-bore WCOT column, however, frequently achieves greater sensitivity than an unsplit injection on a larger-bore column because the lower degree of band broadening may achieve a higher mass/time ratio at the detector (*vide infra*).

Another difficulty is observed with very high boiling components that may form aerosols in the vaporization chamber. Splitters forced to work on aerosols seldom achieve high linearity. German and Horning [22] suggested splitting after the vaporized sample was passed through a short precolumn packed with a low level of SE 30 on a suitable support. This assures that materials are in the vapor phase when they reach the splitting orifice. It also eliminates the contamination of the column with nonvolatile materials that are frequently present in biological samples. Although some degree of band broadening is experienced, in most cases this is not excessive.

When using splitters it is advisable to pass that portion discharged to the atmosphere first through an adsorptive filter—usually activated carbon—and then through the restrictor (Figure 4.10) [23]. This permits room-temperature use of a normal needle valve as the restrictor without the problem of condensation in the valve (which would change the split ratio) and also avoids contamination of the laboratory air with the major portion of the injected materials. This latter should be of concern in all cases, most particularly with chemicals of recognized toxicity.

The split ratio expresses the moles of an injected sample component placed on the column relative to the moles of that sample component shunted to atmosphere. It is normally assumed that the injected sample is completely vaporized and thoroughly blended with the carrier gas stream. The split ratio is sometimes defined as

$$\frac{F_c + F_v}{F_c} \tag{4.1}$$

where F_c is the flow through the column and F_v is the flow through the splitter vent. (Obviously the same volume and time units must be used for both quantities.) More generally, however, it is defined as [24]

$$F_c/F_v \tag{4.2}$$

Except for the fact that the two expressions are precisely opposite (i.e., 100:1 versus 1:100), the difference between (4.1) and (4.2) is significant only at very low split ratios. The latter expression is more generally used and is preferred. These flows can be measured with a suitable bubble flow meter, but not all detectors are designed to facilitate the measurement of carrier gas flow. If the carrier gas flow is to be measured directly, both the hydrogen and make-up gas must be turned off and the pressures in these lines allowed to dissipate. Because the volume of column flow is small, minor leaks within the detector or at the flow meter connection can lead to significant errors.

Alternatively, the volume of carrier gas flow through the column can be calculated and compared to the measured volume discharged through the splitter outlet. If the carrier gas were non-compressible, the flow through the column could be calculated from the relationship

$$F_c = \frac{\pi r^2 L}{t_M} \tag{4.3}$$

where r and L are expressed in centimeters and t_M in minutes. Several factors affect the accuracy of this calculation. F_v is usually measured with a bubble meter, and the vapor pressure of the water makes a contribution so slight that it is normally ignored. Of much greater importance are temperature and pressure (i.e., gas compressibility) correction factors that should be applied to the F_c calculation. The gas compressibility correction factor takes the form

$$j = \frac{3}{2} \frac{p^2 - 1}{p^3 - 1} \tag{4.4}$$

where p equals the ratio of inlet to outlet pressures. Pressure drop correction factors are shown in Table 4.3.

The volume of gas flowing through the column, F_c, corrected to room temperature (25°C) and atmospheric pressure, can also be calculated from

$$F_c = \frac{\bar{u} \pi r^2 298}{T_{col} j} \tag{4.5}$$

where r is the column radius (not diameter) in centimeters, T_{col} represents the absolute column temperature, and j is the pressure

TABLE 4.3

Pressure-Gradient Correction Factors[a]

p	0.00	0.01	0.02	0.03	0.04	0.05	0.06	0.07	0.08	0.09
1.00	(1.0000)	0.9950	0.9901	0.9851	0.9803	0.9754	0.9706	0.9658	0.9611	0.9563
1.10	0.9517	0.9470	0.9424	0.9378	0.9332	0.9287	0.9242	0.9198	0.9154	0.9110
1.20	0.9066	0.9023	0.8980	0.8937	0.8895	0.8852	0.8811	0.8769	0.8728	0.8687
1.30	0.8647	0.8606	0.8566	0.8527	0.8487	0.8448	0.8409	0.8371	0.8333	0.8295
1.40	0.8257	0.8219	0.8182	0.8145	0.8109	0.8072	0.8036	0.8001	0.7965	0.7930
1.50	0.7895	0.7860	0.7825	0.7791	0.7757	0.7723	0.7690	0.7657	0.7624	0.7591
1.60	0.7558	0.7528	0.7494	0.7462	0.7430	0.7399	0.7368	0.7337	0.7306	0.7275
1.70	0.7245	0.7215	0.7185	0.7155	0.7126	0.7097	0.7068	0.7039	0.7010	0.6982
1.80	0.6954	0.6926	0.6898	0.6870	0.6843	0.6815	0.6788	0.6762	0.6735	0.6708
1.90	0.6682	0.6656	0.6630	0.6604	0.6579	0.6553	0.6528	0.6503	0.6478	0.6453
2.00	0.6429	0.6404	0.6380	0.6356	0.6332	0.6308	0.6285	0.6261	0.6238	0.6215
2.10	0.6192	0.6169	0.6146	0.6124	0.6101	0.6079	0.6057	0.6035	0.6013	0.5992
2.20	0.5970	0.5949	0.5928	0.5906	0.5886	0.5865	0.5844	0.5823	0.5803	0.5783
2.30	0.5763	0.5742	0.5723	0.5703	0.5683	0.5664	0.5644	0.5625	0.5606	0.5587
2.40	0.5568	0.5549	0.5530	0.5512	0.5493	0.5475	0.5457	0.5438	0.5420	0.5402
2.50	0.5385	0.5367	0.5349	0.5332	0.5314	0.5297	0.5280	0.5263	0.5246	0.5229
2.60	0.5212	0.5196	0.5179	0.5163	0.5146	0.5130	0.5114	0.5098	0.5082	0.5066
2.70	0.5050	0.5034	0.5019	0.5003	0.4988	0.4972	0.4957	0.4942	0.4927	0.4912
2.80	0.4897	0.4882	0.4867	0.4853	0.4838	0.4824	0.4809	0.4795	0.4781	0.4766
2.90	0.4752	0.4738	0.4724	0.4710	0.4697	0.4683	0.4669	0.4656	0.4642	0.4629
3.00	0.4615	0.4602	0.4589	0.4576	0.4563	0.4550	0.4537	0.4524	0.4511	0.4498

[a] $j = \frac{3}{2}(p^2 - 1)/(p^3 - 1)$, where p = inlet pressure/outlet pressure.

correction factor. The split ratio can be shown to be approximately

$$1 : \frac{0.01 T_{col} j}{r^2 \bar{u} \, \sec_{FMO}} \tag{4.6}$$

where $\sec_{FMO}$ represents the seconds required for displacement of 10 cm^3 at the splitter outlet. Agreement between this method and direct measurement ratios is usually within 1%.

Split injections are used for a variety of applications, because good results are relatively easy to obtain, provided one is interested in major component analysis (i.e., where the components to be detected exist at concentrations of 0.01–10% of the sample to be injected). It is usually unsuited to analysis of minor components such as haloforms in water, although a variety of extraction and other sample preparation techniques or highly sensitive specific detectors (e.g., ECD) are sometimes used to extend the applicability of splitters to trace component analysis.

4.5   On-Column Injection

In each of the injection modes discussed to this point the injected sample is vaporized on a heated surface and transported to the column by the carrier gas. Some sample components may exhibit thermal lability and suffer degradation during this process. An injection in which the condensed sample is placed directly within the unheated column, there to dissolve in the liquid phase comprising the first few theoretical plates, should permit vaporization of the sample at the lowest possible temperature and should in theory eliminate the possibility of discriminatory (or nonlinear) injections. In addition, there is a good probability that dissolved or coinjected air would be stripped from the sample prior to its vaporization. On-column injection has been applied to packed and larger-bore capillary columns, and more recently methods have been suggested for its application to smaller-bore capillary columns [25]. However, a very fine syringe needle (32 gauge) is required to fit inside even a 0.32-mm-i.d. column. These fine needles are too fragile for septum piercing, and other methods have been devised. One promising approach utilizes a septum-free valved inlet similar to that pictured in Figure 4.17 [26]. Grob emphasized that to achieve a narrow band of injected sample the

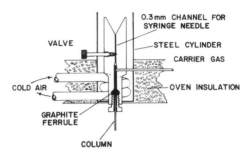

FIGURE 4.17   Schematic of an on-column injector designed for use with 0.32-mm glass capillary columns. (Adapted from Grob [27].)

injection must be made under conditions that permit cold trapping and (more important) the solvent effect to take place. The interrelationships between and the influence of operational conditions on these phenomena have been described [27]. With this particular type of inlet, injections should be made very slowly, and this makes it possible for the temperature of the syringe needle to rise during the course of the injection. This can cause lower-boiling components to vaporize in the needle, leading to sample fractionation and discriminatory injections. Later inlet modifications reportedly correct this difficulty [28].

Another approach to on-column injection was suggested by Rooney [29]. An inlet splitter, similar to those shown in Figures 4.7–4.10, is fitted with a liner possessing a centered constriction (Figure 4.18). Because cold trapping and solvent effects are both important in this injection mode, the initial column temperature should be low; 25°–45°C is usually suitable, depending on the boiling point of the solvent and the sample constituents. With the injector cold and the carrier gas off, a 26-gauge needle is inserted through the septum and left in position. The sample is drawn up into the syringe barrel, and the extra-long 32-gauge needle of the injection syringe is inserted through the 26-gauge guide needle, through the injector assembly, and into the interior of the column. According to Rooney [29] slow injections result in peak splitting with this type of injector, and better results are obtained if the injection is made smoothly and quickly. The syringe and guide needle are then withdrawn, the carrier gas is turned on, and the program is begun.

An even simpler approach can be utilized to ascertain whether on-column techniques are more suitable for a given analysis. The material to be analyzed is first dissolved in a suitable low-boiling solvent, usually along with one or two internal standards. With the column installed and carrier (and detector) gases on, the oven is raised to some suitable temperature above ambient. The oven heaters are then turned off, the oven opened, and the inlet end of the column quickly disconnected and immersed in the sample solution. Solution is drawn into that end of the column as the column cools. The column is reconnected, the oven closed, and the temperature increased. If results indicate a lower degree of sample attrition, modifications to permit a better controlled on-column injection may be in order.

Alternatively, with the carrier gas off and the inlet end of the column disconnected, a micro syringe with a 32 gauge needle can

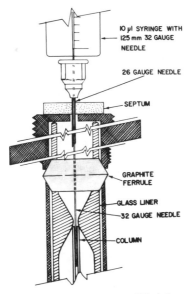

FIGURE 4.18 A standard inlet splitter modified for on-column injection by substituting for the normal glass liner one containing a fine constriction. Sketch condensed longtitudinally by removing a section approximately 1 in. long located just below the septum cap, and another approximately 2 in. long lying just above the graphite ferrule, as indicated. See text for details of operation. (After Rooney [29].)

be used to deposit a controlled amount of sample in the column; care should be exercised to avoid a complete bridging of the column with the sample plug, because there will be a slight reverse flow of make-up gas through the column. If desired, excess solvent can be permitted to evaporate in this low flow stream. The column is then reconnected, carrier gas is turned on and the program begun.

On-column injection techniques are also used with thin-wall large diameter (>0.3 mm i.d.) fused silica columns. When the outer protective coating is removed, such columns have a tendency to break on the slightest provocation. Hence insertion of the syringe needle into the unprotected column bore might be expected to cause problems, but the columns seem to perform quite satisfactorily with on-column injection.

4.6   Other Injection Modes

Several other methods of sample injection are used to a lesser degree with small-bore capillary columns. The direct injection of gas or headspace samples will be considered in Chapter 11. A "falling-needle" technique, which allows the removal of solvent in the analysis of higher-boiling compounds, has been described [30]. The needle is a pointed glass rod that is attached to a short piece of glass-enclosed iron rod. This permits the vertical position of the needle to be controlled by an external magnet. With the needle in the raised position (Figure 4.19, left), several drops of sample solution are placed on the probe tip. The carrier gas stream entering the inlet is split into two flow paths, one through the column and the other over and past the needle and through the outlet restriction to the atmosphere. The needle is held in this position until the solvent has evaporated; the time required for this step will be governed by the inlet temperature, the flow rate of gas, and the volatility of the solvent, and must be determined experimentally. The needle is then lowered into the heated zone (Figure 4.19, right), and the residue vaporized onto column. Specialized pyrolysis techniques, discussed in Chapter 13, have also been used (under lower temperature conditions) to achieve splitless injection of solvent-free samples [31], and valve-switched injections have been explored [32, 33]. Several of the more com-

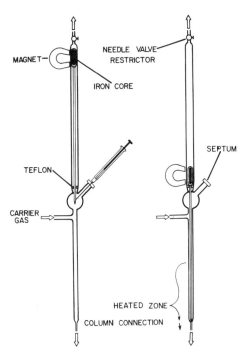

FIGURE 4.19   Falling-needle injector. (Adapted from van der Berg [30].) With the glass needle in the raised position (left), the sample (dissolved in a low boiling solvent) is applied to the needle with a standard syringe. After the solvent has evaporated, the needle is lowered into the heated zone and the sample is vaporized to the column. See text for details.

mon approaches to sample injection have been reviewed by Schomburg *et al.* [34], and a capillary-compatible capsule insertion method of sample injection has been described [35].

4.7   Thermally Focused Inlet

Rijks *et al.* [36] described a novel all-glass inlet that utilized thermal focusing to sharpen the sample plug delivered to the column (Figure 4.20). Air, chilled by passage through a metal coil immersed in a coolant such as dry ice-acetone, is introduced at the coolant inlet, whence it passes over the trapping capillary and vents to atmosphere. This establishes a temperature gradient

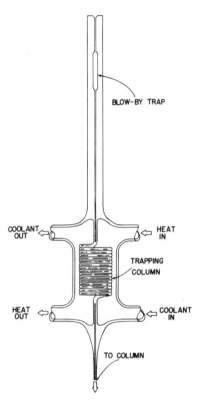

FIGURE 4.20   Thermally focused inlet. (After Rijks *et al.* [37].) The 6-mm-o.d. sample introduction tube is designed for insertion into a ¼-in. heated inlet, via a suitable ferrule (e.g., graphite or Vespel). The blow-by trap, which results from a high carrier gas velocity above and below that area due to the restricted passage around the syringe needle, is a feature that can be included on most inlets. The existence of a temperature gradient during both the trapping and heating steps ensures that the rear of the sample plug is always moving at a higher velocity than is the front of the sample plug; hence band narrowing is achieved. The unit is designed for a Teflon-bridged connection to the capillary column as shown, but is easily adaptable to other connection modes (see Figure 5.6).

along the trapping capillary, which has been coated with liquid phase. Because the sample components encounter continually decreasing temperatures as they proceed through the trapping column, partition ratios continuously increase; the front of the sample plug moves more slowly than the rear of that plug, and the band is narrowed or focused. The coolant is then shut off, and heated air introduced at the top of the trapping section to emerge at the bottom. As the capillary is heated, the rear of the band again moves more rapidly than the front of the band; thermal focusing occurs both during sample trapping and sample heating.

4.8 Syringe Technique

The operator's injection technique can have a dramatic influence on the quality of the resultant chromatogram. Ideally, the sample delivered to the column should accurately represent the injected material, and it should occupy a minimum length of column. Grob and Grob [37] emphasized that the in-needle injection is a poor technique, a generalization that is probably true for all injection modes that utilize syringes, especially for split and splitless injection. When the solution-filled needle is introduced without moving the plunger, the needle gradually heats up, causing the more volatile sample components to distill from the needle. Injections are never total; some material usually remains in the injection syringe. Where volatile components have been distilled off, the remaining material will be richer in the higher boiling components, and the composition of the sample reaching the column is not the same as that of the original sample composition. Injections are more representative if the needle is empty when it is inserted into the inlet. The plunger should be retracted to withdraw the sample into the glass barrel, and the empty needle inserted and allowed to warm up for 3–4 sec in the heated inlet. The injection is then accomplished by thrusting the plunger rapidly home ("hot-needle injection") [37]. Splitters are less forgiving of poor injection technique; hot needle injections via an inlet splitter can result in doublets (Section 14.7), and any materials in the flow path that are capable of retarding a portion of the injected solutes (e.g., by solution, absorption, or adsorption) can lead to defective results. Crumbs of silicone rubber from the septum, scrapings of graphite from ferrules, or high boiling residues from previous injections can absorp a portion of the injection and release it more slowly to give broadened or tailing peaks. Short pieces of coated column broken off and allowed to remain in the inlet splitter can function in this same way; if these lodge in the inlet in such a way that the carrier gas stream is actually divided into two or more paths, peak doublets can result (Sections 14.7 and 16.3).

4.9   Choosing the Injection Mode

From standpoints of ease of instrument conversion, simplicity of the method, (i.e., the probability of a novice obtaining good results) and quantitative reliability, many authorities prefer inlet splitters. Normally the components of interest must exist at levels of 0.01–10% in the injected sample, although methods for heart-cutting and solvent-venting (e.g., [38, 39]) have been suggested to extend this range. Components that exist at levels below 0.01% are usually more amenable to analysis by "splitless" injection (i.e., concentration by the solvent effect), but considerable experimentation may be required to select the proper solvent, amount of solvent (injection size), concentration of components, the initial temperature and the injection and purge times. Splitless injection is not recommended for the analysis of lower-boiling components that elute before the solvent. On-column injection promises to become a very useful technique with particular applicability in the area of samples that are prone to thermal degradation, including many biological compounds, but it can lead to more rapid column degradation. The valved version [26] also offers the advantage of a septum-free inlet. For both splitless and on-column injection, it is difficult to estimate the precise moment of injection; hence t_R values are usually determined by reiterative techniques (Section 7.2). In the final analysis none of these injection methods can be designated as superior for all samples in all cases; the method of sample injection should be selected with a view to the nature of the sample, the results desired, operator skill, and equipment availability.

References

1. **Grob, K.,** and **Grob, G.,** *J. Chromatogr. Sci.* **7,** 584 (1969).
2. **Grob, K.,** and **Grob, G.,** *J. Chromatogr. Sci.* **7,** 587 (1969).
3. **Purcell, J. E., Downs, H. D.,** and **Ettre, L. S.,** *Chromatographia* **8,** 605 (1975).
4. **Schomburg, G.,** and **Husmann, H.,** *Chromatographia* **8,** 517 (1975).

5. **Harris, W. E.,** *J. Chromatogr. Sci.* **11,** 184 (1973).
6. **Grob, K.,** and **Grob, G.,** *Chromatographia* **5,** 3 (1972).
7. **Grob, K.,** and **Grob, K., Jr.,** *J. Chromatogr.* **94,** 53 (1974).
8. **Grob, K.,** and **Grob, K., Jr.,** *HRC&CC* **1,** 57 (1978).
9. **Jennings, W. G., Freeman, R. R.,** and **Rooney, T. A.,** *HRC&CC* **1,** 275 (1978).
10. **Miller, R. A.,** and **Jennings, W.,** *HRC&CC* **2,** 72 (1979).
11. **Schulte, E.,** and **Acker, L.,** *Z. Anal. Chem.* **268,** 260 (1976).
12. **Freeman, R. R.,** personal communication (1979).
13. **Ettre, L. S.,** "Open Tubular Columns in Gas Chromatography." Plenum, New York, 1965.
14. **Halász, I.,** and **Schneider, W.,** *Anal. Chem.* **33,** 979 (1961).
15. **Ettre, L. S.,** and **Averill, W.,** *Anal. Chem.* **33,** 680 (1961).
16. **Condon, R. D.,** and **Ettre, L. S.,** *in* "Instrumentation in Gas Chromatography" (J. Krugers, ed.), p. 87. Centrex Publ., Eindhoven, Netherlands, 1968.
17. **Watanabe, C., Tomita, H.,** and **Sato, N.,** *HRC&CC* **2,** 481 (1979).
18. **Ettre, L. S.,** and **Kabot, F. J.,** *Anal. Chem.* **34,** 1431 (1962).
19. **Jennings, W. G.,** *J. Chromatogr. Sci.* **13,** 185 (1975).
20. **Hartigan, M. J.,** and **Ettre, L. S.,** *J. Chromatogr.* **119,** 187 (1976).
21. **Jennings, W. G.,** *J. Food Chem.* **2,** 185 (1977).
22. **German, A. L.,** and **Horning, E. C.,** *Anal. Lett.* **5,** 619 (1972).
23. **Jennings, W. G.,** and **Adam, S.,** *Anal. Biochem.* **69,** 61 (1975).
24. **Ettre, L. S.,** and **Averill, W.,** *Anal. Chem.* **33,** 680 (1961).
25. **Schomburg, G., Diehlmann, R., Husmann, H.,** and **Weeke, F.,** *Chromatographia* **10,** 383 (1977).
26. **Grob, K.,** and **Grob, K., Jr.,** *J. Chromatogr.* **151,** 311 (1978).
27. **Grob, K.,** *HRC&CC* **1,** 263 (1978).
28. **Galli, M., Trestianu, S.,** and **Grob, K., Jr.,** *HRC&CC* **2,** 366 (1979).
29. **Rooney, T. A.,** personal communication (1979).
30. **van der Berg, M. L. J.,** Doctoral Thesis, Tech. Univ., Delft, Netherlands, 1975.
31. **deLeeuw, J. S., Maters, W. L., van der Meent, D.,** and **Boon, J. J.,** *Anal. Chem.* **49,** 1881 (1977).
32. **Langlais, R., Schlenkermann, R.,** and **Weinberg, M.,** *Chromatographia* **9,** 601 (1976).
33. **Gasper, G., Arpino, P.,** and **Guichon, G.,** *J. Chromatogr. Sci.* **15,** 256 (1977).
34. **Schomburg, G., Behlau, H., Diehlmann, R., Weeke, F.,** and **Husmann, H.,** *J. Chromatogr.* **142,** 87 (1977).
35. **Buser, H. H.,** and **Widmer, H. M.,** *HRC&CC* **2,** 177 (1979).
36. **Rijks, J. A., Drozd, J.,** and **Novak, J.,** *in* "Advances in Chromatography, 1979" (A. Zlatkis, ed.), p. 195. Chromatogr. Symp., Univ. of Houston, Houston, Texas, 1979.
37. **Grob, K.,** and **Grob, G.,** *HRC&CC* **2,** 109 (1979).
38. **Anderson, E. L.,** and **Bertsch, W.,** *HRC&CC* **1,** 13 (1978).
39. **Miller, R. J., Stearns, S. D.,** and **Freeman, R. A.,** *HRC&CC* **2,** 55 (1979).

CHAPTER 5

COLUMN INSTALLATION

5.1 General Considerations

Most difficulties experienced with glass capillary GC relate to errors committed by the user during the column installation step. Problems such as much lower theoretical plate numbers than the manufacturer's specification, broad or tailing peaks, adsorptive tailing or the subtraction of active compounds, decreased sensitivity or poor separation may result from the improper installation of a good column in a well-designed system. The type of fittings used for column attachment should be selected with a view to the type of compounds to be analyzed, the anticipated operating temperatures, the frequency with which the column will be changed, and (at the inlet end) the injection mode, i.e., split, splitless, or on-column.

5.2 Column Attachment; Inlet

As discussed in the preceding chapter, inlet splitters require that the split point be located in a high-velocity gas zone for ideal performance. Three general methods of attaching the glass capillary column to an inlet splitter are in common use. The column end can be straightened so that it extends through the point of attachment to the split point in the inlet; the unstraightened end of the glass capillary column can be coupled—usually with some version of a "zero-dead-volume" fitting—to a straight extension that projects to the split point in the inlet; specially relieved fittings, capable of accepting the curvature of the unstraightened column end, can be utilized so that the column end projects only slightly into the inlet but terminates at the split point (i.e, in a high-velocity gas zone). With splitless injection it is preferable that the columns not project into the inlet, because this would create unswept areas during the injection step. Instead the column is usually attached directly to the inlet extremity. These various configurations are represented schematically in Figure 5.1.

5.3   Column Straightening

Straightening the ends of glass capillary columns requires no special manual dexterity, but practice does help. The column

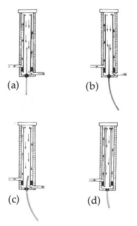

FIGURE 5.1 Schematics of column mountings. (a) Split mode, straightened column terminating in high-velocity gas zone, split point locatable within heated zone of injector. Generally applicable for all samples and preferred for samples with high boiling points (e.g., crude oils). (b) Split mode, unstraightened column with butt joint in low-velocity gas zone, split point within heated zone of injector. (c) Split mode, unstraightened column terminating in high-velocity gas zone, split point in oven area. Suitable for all but highest boiling samples. (d) Preferred configuration for splitless injection.

should first be dispersed along a horizontal rod. This also offers an opportunity to count the coils and determine the column length; if necessary, the column should be untangled at this time. For most applications column ends should be straightened under gas flow, because liquid phase decomposition products are adsorptive. To accomplish this, one end of the column is connected to a low-pressure gas supply; nitrogen or helium should be used for most columns, but air seems to give better results for columns coated with SP 1000 or SP 2250. The column is carefully oriented and a small soft flame played over a length of column so that it straightens of its own volition under the influence of gravity (Figure 5.2). Great care must be taken to ensure that the flame is not inadvertently directed at other areas of the column. The amount of heat required at any given point is related to the mass of the column end hanging from that point; the application of too much heat when there is a considerable mass of hanging column may cause the column end to break away or may lead to a localized

Figure 5.2 Flame-straightening column ends. A low-velocity gas flow is usually advisable; avoid heating other portions of column. Using minimum heat, allow the column to straighten under the impetus of gravity; frequent reorientation of the column will help.

change in column diameter. Flames fed by high-velocity gas (i.e., a propane torch) must be used with great care because they also exert a side thrust on the softened column. The limited mass of the final centimeter of column is rarely large enough that the force of gravity can pull it straight without excessive heat, and it is usually wiser to score the end with a glass knife and remove this section. Whenever a column is trimmed, it should be held so that any glass shards fall out of and not into the column as it is broken. The break will be more nearly square if it is achieved by applying tension at either side of the score rather than bending at that point.

Columns of soda-lime or soft glass (this would include practically all etched columns) will require a softer flame than columns of borosilicate glass. Although the novice finds it simpler to produce straighter column ends by beginning at the end and proceeding toward the center, this may encourage the movement of liquid phase toward the center and result in a column block. It is usually wiser to begin straightening 10–12 cm from the end and work toward the end; certainly care must be taken not to overheat the column. It should then be examined under low-power magnification to ensure that no constrictions have resulted and that the liquid phase has not moved under intense localized heat to form a column blockage. Usually some liquid phase deterioration results from this treatment, but it would not appear to cause serious consequences.

Columns that have been flame-straightened occasionally exhibit tailing with some compounds, which in some cases appears to be due to the heat-induced exposure of active adsorptive sites on the glass, and in cases where straightening was accomplished without the benefit of gas flow, may relate to the adsorptive properties of decomposition products of the liquid phase.

Tailing problems of this type can sometimes be corrected by temporarily attaching one straightened end of the column to the inlet and adjusting the inlet pressure to 0.5–1 atm. A 1-2-ml syringe containing a solution of 5–10 mg/ml Carbowax 20 M in dichlormethane is attached to the other end, and against the gas pressure the solution is forced into the column as a single coherent plug. The solution should fill that section of the column that has been straightened, but it should go no further. The pressure in the syringe is then released gradually so that the Carbowax 20 M solution is forced out by the gas pressure at a velocity of ~1 cm/sec, and the gas flow is continued for ~1 hr. The column is reversed, and the process is repeated on the other end. Carbowax 20 M is reasonably surface active and should adsorb to and deactivate any adsorptive sites. Even on highly nonpolar columns such as SE 30 or OV 101, the effect on retention behavior is not significant. The passage of Carbowax 20 M decomposition products through the column [1] is also widely used to correct these problems. This method is discussed in Section 5.5. Injection of triethanol amine [2] or tri-isopropyl amine [3] is a simple method of temporarily lessening—but rarely eliminating—the activity of adsorptive sites, but some users report that columns deactivated in this way may then abstract aldehydes.

Column ends can also be straightened with a short length (~5 cm) of ⅛-in. stainless steel tubing heated electrically to a point where the interior of the tube exhibits a dull red glow. In our own laboratory, we use a dental brazing transformer, supplying ~100 A at ~1 V, to heat the tube, although it could also be lagged directly or heated with a flame. Lubrication of the straightening tube with powdered graphite or molybdenum disulfide reduces the incidence of breakage. (The latter is available in spray cans for the lubrication of motorcycle chains.) The end of the column is fed through the straightening tube and immediately withdrawn (Figure 5.3). Movement or decomposition of the liquid phase occurs much less frequently with the heated-tube straightening devices.

One of the major advantages offered by the very fine bore thin wall flexible silica columns [4] described in Chapter 2 is that they are inherently straight; the desired length of column is merely pulled from the coil ties. The very thin wall necessary to impart flexibility to these columns results in a smaller outside diameter. Consequently, it is sometimes possible to thrust the outlet end of the column well up inside the flame jet, so that the column terminates immediately below the flame itself. This sometimes eliminates a small degree of peak tailing that persists when columns are connected in a normal manner, which is apparently due to active sites in the flame jet itself.

Some workers have recommended that short sections of glass (Section 5.5), glass-lined stainless steel or platinum iridium tubing be more or less permanently attached to the inlet and detector

Figure 5.3   Straightening column ends with an electrically heated straightening tube. Power supply is a dental brazing transformer, although an insulated tube lagged with nichrome wire works equally well.

fittings of the chromatograph, and that heat-shrink Teflon tubing be used to attach the column to these projections; it would be difficult to advise too strongly against this practice. Not only does it involve two butt joints in low-velocity gas zones with their associated problems, but even minute traces of Teflon exposed to the flow stream can lead to adsorptive tailing. There is also some evidence that the Teflon connection is oxygen-permeable, leading to more rapid column deterioration [5]. It is a simple matter to change columns in such a system, but one usually finds that increased convenience was purchased at the high price of inferior chromatograms.

5.4   Column Attachment; Detector

After the column has been connected to the inlet and the carrier gas turned on, it is usually wise to demonstrate carrier gas flow through the column while the outlet end of the column is still free. The free end of the column can be immersed in distilled water, or a forefinger moistened with saliva can be used to verify carrier gas flow (Figure 5.4). Contamination of this end of the column with leak detector fluids or other substances that may contribute to detector noise should be avoided.

The column end should then be inserted into the detector fitting as far as possible, and certainly beyond the point of hydrogen and/or make-up gas addition, so that the column terminates in a high-velocity gas zone. Figure 5.5 is a schematic representation of the gas velocity through various areas of a detector connection that utilizes a make-up gas adaptor (Section 14.4). If the end of the column is located between the column connection point and the point of make-up gas introducton, it lies in the zone of lowest gas velocity. The volume of gas through this zone is restricted to that of the carrier gas, 1-2 cm³/min. (Termination of the column in this zone may happen inadvertently owing to breakage.) This results in tailing and remixing of separated peaks.

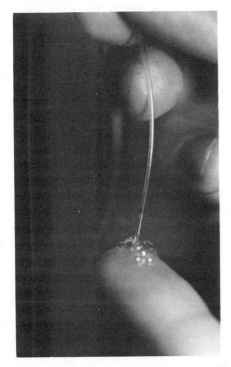

Figure 5.4 Verification of column flow prior to connecting detector end is usually wise.

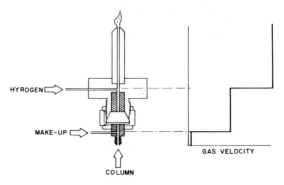

HYROGEN

MAKE-UP

GAS VELOCITY

COLUMN

Figure 5.5 Schematic of a flame ionization detector with make-up gas adaptor (left), and (right) the relative gas velocity at different positions through that connection. See text for discussion.

If the column terminates above the point of make-up gas introduction (if used), it lies in a zone of much higher gas velocity and eluting fractions are caught up and swept to detection much more rapidly. The ideal column termination point is in the zone of highest gas velocity, i.e., above the point of hydrogen introduction (see also Figure 14.6).

5.5   Column Connection Fittings

Figure 5.6 illustrates several common methods of attaching capillary columns. The connection at the detector end should, of course, have provision for make-up and/or hydrogen addition at some point before the termination of the column, as discussed in Section 5.4; for split injection the inlet end of the column should also terminate in a high-velocity gas zone. For some other injection modes (e.g., splitless), the left example in Figure 5.6 can be regarded as minimally acceptable for nondemanding applications. Although it offers the lure of simplicity and convenience, the Teflon connection becomes unreliable at about 250°C, and failure should be expected at about 280°C. In addition, the existence of a butt joint in a low-velocity zone and the exposure of even a minute amount of the Teflon elastomer to the flow stream cause problems. The center example of Figure 5.6 uses a standard fitting but requires straightened column ends; an adaptation of this is used in the inlet shown in Figures 4.7 and 4.10, and the make-up gas adaptor illustrated in Figure 14.6. The example on the right of Figure 5.6 uses a special fitting but allows the use of unstraightened columns. The make-up gas adaptor shown in Figure 14.7 is based on this principle.

Silicone rubber septa, lead, Teflon, Vespel, and graphite have all been used as ferrules for the attachment of the column and other glass components of the system. Silicone rubber septa, 3.5–4-mm diameter, can be readily prepared with a suitable cork borer. They are satisfactory for temperatures up to about 200°, but beyond this their elasticity is relatively short lived. Also, when the fitting is being secured or tightened, septa-type connections have a tendency to twist and shift, which often causes the column to shear at this critical point.

When ferrules of lead or virgin Vespel are used, they tend to fuse to the column and usually cause breakage on removal. This is not a tragedy as far as the column is concerned; small differences in column length will have no discernible effect on its efficiency, but the ferrules can rarely be reused. A graphitized Vespel seems more satisfactory, but the limited compressibility of the Vespels requires close tolerances between the diameter of the column and the hole through the ferrule. Grob described the properties of

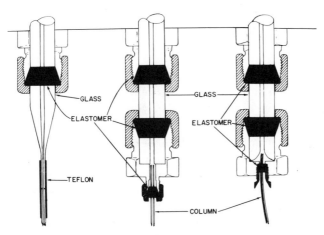

Figure 5.6 Some commonly used methods of column attachment. See text for details.

Kalrez, a Teflon-like elastomer produced by DuPont, as they relate to column mounting. His drawings indicate that the material continues to form in a process analogous to cold flow for some time during use [6].

The normal graphite ferrules suffer from brittleness and frequently break on removal; a more flexible graphite ferrule is now available at reasonable cost. This will withstand temperatures as high as 450°C and is capable of considerable compression, so that tolerances are not critical; the bore can be reexpanded or enlarged by judicious application of a polished taper, such as a darning needle. Provided it is not abused, particularly by overtightening, the graphite ferrule can be reused many times (Figure 5.7). When it is used with normal fittings [Figure 5.7(2), (3)] the ferrule may lodge in the nut on removal; it is normally allowed to remain there and the nut and ferrule are reused as a unit. With the relieved fitting and reversed ferrule [Figure 5.7(1)] the ferrule is best recovered by releasing the fitting and moving the column

Figure 5.7 Graphite ferrules, used without back ferrules or back-up washers, provide an excellent seal; with careful handling they can be reused many times. In order from left to right (1) a reversed 1-mm ferrule can be used with a relieved fitting on either straightened or unstraightened columns (e.g., Figure 4.12); (2) normal $\frac{1}{16}$-in. fitting used with a 1-mm graphite ferrule (e.g., Figures 4.9 and 4.18); (3) standard 6-mm graphite ferrule used with standard fitting; (4) double tapered graphite ferrule used with inlets similar to those in Figure 4.9; splitter housing shown reversed to expose the inner tapered seal.

very slightly; the gas pressure will then usually expel the column with the ferrule. In extreme cases where the fitting has been seated too tightly, recovery of the reversed ferrule may require careful insertion of the tapered end of a very fine rattail file into the ferrule bore. This is seated with a slight twist, and the ferrule is withdrawn for reuse. Schomburg *et al.* [7] have described a sleeved graphite ferrule. Graphite, Vespel, and most other ferrules have a tendency to bake onto the protective coatings used with fused silica or fused quartz columns (Section 2.2). In such cases it is probably advisable to leave the captive ferrule and nut attached to and reuse it as an integral part of that column.

Great care should be exercised to ensure that none of these materials—Teflon, silicone rubber, Vespel, or graphite—comes into contact with the flow stream. If this happens, tailing quickly becomes evident. Pieces of broken column that are allowed to remain in the flow path can lead to tailing, and if they lodge in the inlet splitter in such a way that they pass a portion of the flow stream, doublet peaks can be observed.

After both column attachment fittings have been installed finger tight, the system should be pressurized and checked for leaks. A leaky fitting should be tightened just enough to eliminate the leak; overtightening will ruin most ferrules and may break the column. An electronic leak detector is preferred, but leak detector solutions can also be used, provided they are not incompatible with the detection mode to be utilized (e.g., phosphate-containing detergents may lead to severe noise problems with N/P detectors). A minimum quantity of solution should be applied and the excess removed by blotting to minimize the possibility of its slow aspiration into the system.

5.6 Initial Evaluation

Hydrocarbon mixtures can be used to good advantage in evaluating the system, because they should be relatively immune to adsorptive interactions that cause tailing. For maximum separation efficiency the optimum average linear carrier gas velocity should be used; values of approximately 25 cm/sec for helium carrier gas or 40 cm/sec for hydrogen carrier gas are reasonable initial values. It is probable that much higher velocities will be employed for the actual analyses, as it is usually beneficial to trade surplus separation power for shorter analysis times (Chapter 8). Commercial columns are usually supplied with a test chromatogram, specifying the temperature, carrier gas velocity, and test compounds. It is normally wise to attempt to duplicate that test run.

The first test is usually a methane injection, which can also be used to set the flow rate through the column.

$$L/t_M = \bar{u}$$

where L is expressed in centimeters and t_M in seconds. Chart speeds of at least 2 cm/min should be employed during testing so that peaks can be readily evaluated for asymmetry or tailing. Figure 5.8 shows typical faults, and additional information will be found in Chapters 14 and 16. The methane source and suitable substitutes are discussed in Section 14.2.

Methane should produce a needle-sharp peak; if this is not the case, the problem must be diagnosed and rectified. The most probable causes are dead volume at the inlet or detector end. With an inlet splitter, the split ratio may be too low, or for some other reason the injection chamber may not have been swept clean. This can be associated with a misplaced, omitted, or broken in-

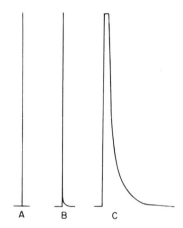

A   B   C

Figure 5.8 Results of the methane injection. (a) ideal peak, (b) tailing (see text), (c) severe tailing (see text).

jection port liner, resulting in dead pockets whose volume slowly diffuses into the carrier gas stream. In any of these cases, the column would receive a long continuous injection, resulting in a broad or asymmetrically distorted peak even for methane (Figure 5.8c). Overtightening the column nut at the detector end may have caused breakage within the ferrule, or lateral displacement of a silicone rubber mounting septum, if used, may have sheared the column. When this happens, the column effluent is subject to remixing before it slowly diffuses into the stream of make-up gas, again resulting in broader peaks. If the defect is less severe—a very small dead space in the inlet, or the outlet of the column terminating just 1 or 2 mm below the carrier gas inlet—the peak may exhibit good conformation but possess a slight tail (Figure 5.8b).

Once good results are achieved with a methane injection, a hydrocarbon series whose higher homologs have boiling points slightly in excess of the compounds to be investigated should be injected; inclusion of one or two alcohols in the test mixture can aid in diagnosis of adsorptive problems. Severe broadening or asymmetry in later peaks may relate to insufficient temperature at the inlet or detector or to the need to preheat the make-up gas (see Chapter 14). Asymmetric tailing in all peaks will be observed when the inlet or front of the column becomes contaminated with a material in which the sample can dissolve. Crumbs of silicone rubber from septa that may fall into the inlet or become lodged in the column, or globules of liquid phase within the column cause defects of this type. If the hydrocarbon peaks exhibit good conformation but an alcohol such as hexanol-1 tails, adsorption is indicated. This can sometimes be corrected by silylation of the inlet surfaces, but the procedure described by de Nijs et al. [1] is generally more satisfactory. As used by these workers, the method involves the insertion of a $\frac{1}{4}$-in.-o.d. glass tube packed with 5% Carbowax 20 M on Chromosorb W AW into the heated injection port of a gas chromatograph. The end of that tube projecting into the oven compartment is drawn out to match approximately the diameter of the capillary column to be deactivated and is connected to the column with heat-shrinkable Teflon tubing (similar to Figure 5.6). Alternatively, a standard $\frac{1}{4}$- $\times$ $\frac{1}{16}$-in. reducing union can be used to connect the capillary column to the bottom of the packed tube. The temperature of the "precolumn" is maintained at ~260°C and that of the column at ~250° (precise temperatures do not seem to be important); vapors from the precolumn are allowed to bleed through the capillary for several hours (overnight) at a carrier gas flow of ~3 cm³/min (see Section 10.6, Figure 10.4).

5.7   Test Mixtures

Results from the injection of test mixtures can provide several types of useful information. Not only can they be used to measure the efficiency of the system in terms of theoretical plates or Trennzahl (Chapter 6), but they can be used to monitor the loss of liquid phase as evidenced by a decrease in partition ratios, to diagnose liquid phase deterioration leading to polarity changes and shifts in the relative retentions of compounds possessing dissimilar functional groups, to indicate the need for column deactivation, and to demonstrate the suitability (or unsuitability) of the system for the analysis of certain compounds.

Grob and Grob [8] suggested a polarity test mixture that has wide applicability. This is composed of 5-nonanone, 1-octanol, and naphthalene, each at the level of 0.1% in hexane, together with four n-paraffin hydrocarbons; the latter are varied, depending on the polarity of the column to be tested. The least polar columns (methyl silicones) require a Polarity 12 test solution, in which the hydrocarbons are C_9-C_{12}. Highly polar liquid phases require a Polarity 16 test solution, containing 5-nonanone, 1-octanol, naphthalene, and $C_{13}-C_{16}$ hydrocarbons. This ensures that the critical test compounds are dispersed among the appropriate n-paraffin hydrocarbons, so that their retention indices can be calculated (Chapter 7). Carbowax 20 M can be evaluated with a Polarity 14 solution (i.e., hydrocarbons $C_{11}-C_{14}$). (The use of these solutions to evaluate column deterioration will be discussed in Chapter 10.)

Evaluations should utilize small injections and high instrument sensitivity; hydrocarbon peaks should be sharp and symmetrical with no evidence of overloading. Overloads are characterized by an asymmetric peak exhibiting a "leading tail" (Chapter 16), which can be diagnosed with greater sensitivity by comparing the time required for the up and down strokes of the recorder pen. If the hydrocarbon peaks are well formed, attention can then be directed to the more critical test substances. Asymmetry, tailing, or abstraction of naphthalene may indicate the presence of metal in the flow stream (e.g., a cracked inlet liner). Octanol will quickly point up any adsorptive sites by exhibiting severe tailing, and in extreme cases the octanol peak may disappear. A minor degree of alcohol tailing does not necessarily condemn the column, although if the samples to be analyzed do contain alcohols, it may indicate the need for a deactivation treatment. If the samples to be analyzed are petroleum hydrocarbons, then even moderate alcohol tailing is probably of little or no concern. Once the ketone begins to exhibit tailing, it indicates that the system is becoming too active and some corrective action should be taken.

Grob also suggested the use of an acidity test mixture, containing 1 mg/cm³ of 2,6-dimethylanaline (DMA) and 2,6-dimethylphenol (DMP) in hexane. Injection parameters are set up so that approximately 0.015 μl of solution, containing some 15 ng each of DMA and DMP, is introduced onto the column at a column temperature of 130°C. With a neutral column, both peaks should be symmetrical and of equal size. Results of Dandeneau and Zerenner [4] indicate that other phenols might provide a more demanding test; columns prepared from soda-lime glass produced a well formed peak for 2,6-DMP, but showed severe distortion and/or abstraction on a variety of other phenols (Figure 2.6). There are reports that 2,6-DMA can produce anomolous results on fused silica columns coated with Carbowax 20 M, and the above authors used 3,4-DMP and dicyclohexylamine in their evaluations of fused silica columns.

While these can be useful tests, they are sometimes misapplied, in that a neutral column is not always desired. The FFAP or SP 1000 liquid phases are intentionally acidic, designed to pass acidic solutes; it should not be too surprising to find that they produce severe asymmetry on a DMA peak. Nicotine is also a useful test substance; it has been the author's experience that some columns that give a neutral test with the acidity test mixture are incapable of passing nicotine.

More recently, a very demanding test mixture has been proposed [9]; the composition of this is shown in Table 5.1. The authors suggest the use of a temperature-programmed test, and evaluation of the loss of specific solutes by a "100% line" drawn

TABLE 5.1
Test Mixture Composition[a]

Component	mg/l	Component	mg/l
Methyl dodecanoate	41.3	Nonanal	40
Methyl undecanoate	41.9	2,3-Butanediol	53
Methyl decanoate	42.3	2,6-Dimethylanaline	32
Decane	28.3	2,6-Dimethylphenol	32
Undecane	28.7	Dicyclohexylamine	31.3
1-Octanol	35.5	2-Ethylhexanoic acid	38

[a] From Grob et al. [9].

through the maxima of nonabsorbed peaks (Figure 5.9). Very few (if any) columns would exhibit 100% passage of the wide range of solutes embraced by this test mixture, but it would permit a meaningful assessment of the suitability or unsuitability of that column for a wide variety of analyses.

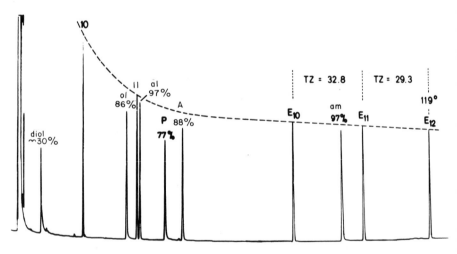

Figure 5.9 Test chromatogram of the demanding test mixture shown in Table 5.1. The dotted line is the 100% line (see text). (From Grob et al. [9].)

References

1. **de Nijs, R. C. M., Franken, J. J., Dooper, R. P. M., and Rijks, J. A.,** *J. Chromatogr.* **167,** 231 (1978).
2. **Sandra, P., and Verzele, M.,** *Chromatographia* **10,** 419 (1977).
3. **Verzele, M., and Sandra, P.,** *HRC&CC* **2,** 303 (1979).
4. **Dandeneau, R. D., and Zerenner, E. H.,** *HRC&CC* **2,** 351 (1979).
5. **Grob, K.,** *J. Chromatogr.* **168,** 563 (1979).
6. **Grob, K.,** *HRC&CC* **1,** 103 (1978).
7. **Schomburg, G., Diehlmann, R., Borwitzky, H., and Husmann, H.,** *J. Chromatogr.* **167,** 337 (1978).
8. **Grob, K., and Grob, G.,** *Chromatographia* **4,** 422 (1971).
9. **Grob, K., Jr., Grob, G., and Grob, K.,** *J. Chromatogr.* **156,** 1 (1978).

CHAPTER 6

MEASURING COLUMN EFFICIENCY

6.1 General Considerations

It would be difficult to overemphasize the fact that most methods for the measurement of column efficiency reflect the quality not only of the column but of the entire system, including the efficiency of the injection process. The inlet, column connections, suitability of the point of auxiliary gas introduction, imperfections in flow characteristics that may be traced to dead volumes, cracked inlet liners, or the existence of transfer lines—in short, from the moment of sample injection until the recorder pen has finished with that peak—anything that affects band broadening or separation or leads to remixing of separated components must affect the column efficiency measurement.

Although most workers recognize the imperfections of the theoretical plate concept (*vide infra*), it is still widely used in reporting column effectiveness and measuring column efficiencies. Columns are usually supplied in specified lengths, sometimes guaranteed to possess a certain number of theoretical (or by a very few suppliers, *effective* theoretical) plates. The number of theoretical plates does give the prospective purchaser some information; usually a column with a large number of theoretical plates is capable of a higher degree of separation than a column with a smaller number of theoretical plates. However, both the functional group and the partition ratio k of the test compound can have an effect on these numbers (see Figure 1.3). Plate numbers (and other evaluation measurements that relate to these) can also be affected by the length of the sample plug injected [1] or manipulated by making the measurement on a small peak that follows a larger peak in which it has experienced a solvent effect [2]. One supplier offers columns with a specified number of effective theoretical plates, without reference to length. Others specify a length of column, with no reference to N or H, which tells us nothing about the powers of separation possessed by that column; longer columns merely require longer analysis times.

There are good arguments for using the height equivalent to a theoretical plate or an effective theoretical plate (h or H) as a measure of column efficiency. If the values are determined at the optimum average linear gas velocity, these values are independent (or very nearly so) of column length, so that column efficiencies can be compared, provided that the other basic column and operational parameters are known. These include column diameter and the partition ratio of the test compound (see Figure 1.3 and Table 6.1).

Ideally, we strive for the necessary number of theoretical plates in the shortest possible length of column (see Chapter 8), i.e., columns of high coating efficiency. Column efficiencies are sometimes reported as the number of theoretical plates per meter, and these values have been used as a criterion of the efficiency of the

TABLE 6.1

r_0 (mm)	h_{min}	n_{max}/m	H_{min}	N_{max}/m
0.05	0.08	12,500	0.13	7700
0.10	0.16	6250	0.26	3850
0.125	0.20	5000	0.32	3125
0.25	0.41	2440	0.64	1560
0.375	0.61	1640	0.96	1040
0.50	0.82	1220	1.28	780

coating operation. Golay [3] gives the relationship

$$h_{min} = r_0 \left[\frac{1 + 6k + 11k^2}{3(1 + k)^2} \right]^{1/2} \tag{6.1}$$

for the maximum theoretical attainable efficiency in WCOT columns. According to this, a perfect coating (assuming $k = 4$) would achieve efficiencies as reflected in Table 6.1. In general, these can be regarded as goals not easily attained, but it is normally observed that nonpolar liquid phases approach these values more readily than do polar liquid phases. Equation (6.1) assumes that the C_L term of the Golay equation [see Eq. (8.1)] is negligibly small; it is applicable only to smooth-bore open tubular columns with uniform films. It should not be applied to PLOT, SCOT, or whisker columns; the validity of its use with deeply etched WCOT columns is questionable.

Although objections have been raised against the use of coating efficiencies for column evaluation, they do offer some distinct advantages. Because they are independent of column length and include column diameter and partition ratio factors, the quality of one column can be realistically compared with that of another. On the other hand, they give, by themselves, no indication as to either the total separation capability or the sample handling capacity of that column. Compare, for example, two 50-m columns, one with an i.d. of 0.20 mm and one with an i.d. of 0.32 mm, both of which exhibit 85% coating efficiency. Intuitively, we recognize that the smaller-diameter column could produce superior analytical chromatograms but that it has a limited sample capacity; because of its greater sample capacity, the larger-diameter column would probably be preferred for GC/MS application. In this case, however, the column diameter has been specified, and by applying Eq. (6.1) (or the data shown in Table 6.1), we can calculate that the larger-bore column can generate about 162,000 theoretical plates on a test compound with a partition ratio of 4.0. The small-bore column, however, is capable of some 262,000 theoretical plates under these same conditions.

The problem with most methods of reporting column efficiencies lies in the implication that the values are additive; if a 20-m column possessing 3000 plates/m exhibits 60,000 theoretical plates, a 60-m column with the same coating efficiency would possess 180,000 theoretical plates. This assumption is probably valid at the optimum average linear carrier gas velocity [4], but usually we elect to use much higher velocities. At carrier gas velocities above $\bar{u}_{opt}$, the gain in plate numbers obtained by resorting to a longer column is governed by the slope of the van Deemter curve, and longer columns exhibit curves with steeper slopes (i.e., they are characterized by larger increases in h per unit increase in $\bar{u}$). Figure 6.1 helps clarify this point. Assume two segments taken from the same column, identical in all respects except length, one being 15 and the other 30 m long. Both columns exhibit an h_{min} of 0.2 mm at $\bar{u}_{opt}$, i.e., theoretical plate numbers of 75,000 and 150,000, respectively. At a carrier gas velocity $\bar{u}_2$, the 15-m column may possess an h of 0.3 mm; because its van Deemter curve has a steeper slope, the 30-m column must suffer a greater increase in h at this same carrier gas velocity. In this hypothetical example we have assumed $h = 0.4$ mm. At the carrier gas velocity $\bar{u}_2$ then, the theoretical plate numbers become 50,000 and 75,000, respectively. Hence doubling the column length at $\bar{u}_{opt}$ would double the number of theoretical plates realized, but doubling the column length at $\bar{u}_2$ would, in this example, yield only a 50% increase in the number of theoretical plates. The faster the carrier gas velocity (provided it exceeds $\bar{u}_{opt}$), the greater this discrepancy because

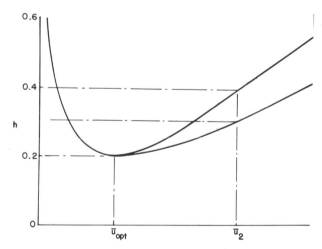

FIGURE 6.1 Hypothetical van Deemter curves for two segments of the same column, where $L_1 = 2L_2$.

longer columns lose efficiency more rapidly per unit increase in $\bar{u}$ than do short columns. Grob and Grob [5] have demonstrated quite effectively that each half of a column exhibits considerably more than one-half of the theoretical plates possessed by the intact column when both are operated at approximately $2\bar{u}_{opt}$.

6.2 Separation Number, Trennzahl (TZ)

Kaiser [6] discussed the imperfections of the preceding methods and stressed the advantages of using the Trennzahl (TZ) or separation number. This method, which is also applicable to programmed temperature determinations, is in some ways a more realistic measure of the separation efficiency of a column, and the column can be evaluated using test compounds of the type for which it will be used. In practice, the separation between two members of a homologous series, differing by one CH_2 unit, is measured (Figure 6.2):

$$TZ = \frac{t_{R(C_{n+1})} - t_{R(C_n)}}{w_{0.5(C_n)} + w_{0.5(C_{n+1})}} - 1 \qquad (6.2)$$

Because the separation number varies with the partition ratios of the test compounds these should also be specified.

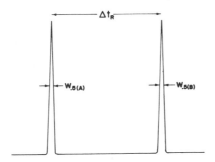

FIGURE 6.2 Method of calculating the separation number

$$TZ = \frac{\Delta t_R}{w_{0.5(A)} + w_{0.5(B)}} - 1$$

where A and B are two members of a homologous series differing by one methylene unit.

Temperature also plays a role, over and above the effect it has on partition ratios. Figure 6.3 shows plots of the Trennzahl as a function of the partition ratio for a series of n-paraffin hydrocarbons at several temperatures. All runs utilized helium carrier gas at an average linear velocity of 25 cm/sec. The data indicate that, under these test conditions, TZ varies inversely with temperature even at constant partition ratio. This relates to the fact that, for a given increase in column temperature, the adjusted retention time of a larger member of a homologous series decreases proportionately more than does the retention time of a smaller homolog; in other words, relative retentions vary inversely with column temperature [7, 8]. As the temperature is lowered, relative retentions become larger and the peak maxima are separated by a greater distance, but a point is eventually reached where diffusion processes are so inhibited [9] that peak shape and resolution suffer. This implies the existence of an optimum temperature for the separation of any two substances on a given column, a subject which was explored by Scott [10] (Section 8.7). In most cases these temperatures are lower than are normally employed and may

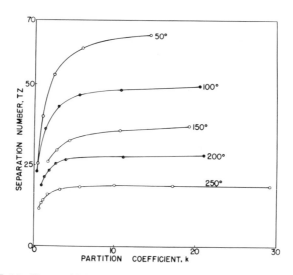

FIGURE 6.3 Trennzahl (separation number) for a series of n-paraffin hydrocarbons at different temperatures on SP 2100 as functions of the partition ratios. Note that at constant partition ratio lower temperatures lead to higher values for TZ within the temperature range investigated.

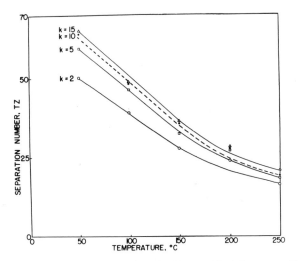

FIGURE 6.4   Effect of temperature on the Trennzahl of the *n*-paraffin hydro-carbons on SP 2100. From the data shown in Figure 6.3.

result in inordinately long analysis times. Figure 6.4 illustrates the effect of column temperature on the separation number, within the range of normally used column temperatures.

The effective peak number (EPN) suggested by Hurrell and Perry [11] also expresses efficiency as the degree of separation achieved on two consecutive members of a homologous series, but at a different degree of resolution:

$$EPN = \frac{2(t_{R(B)} - t_{R(A)})}{w_{(A)} + w_{(B)}} - 1 \qquad (6.3)$$

The two systems have been compared and discussed by Ettre [12], who points out that because $w = 1.699w_{0.5}$ the EPN is defined at a resolution of 4.0σ (where σ is the standard deviation of a Gaussian peak) and the TZ at a resolution of 4.7σ. Hence

$$EPN = 1.17TZ - 0.177 \qquad (6.4)$$

and

$$TZ = 0.85EPN - 0.15 \qquad (6.5)$$

Consecutive homologs are not always readily available, and half the values obtained on pairs differing by two methylene units have sometimes been used for evaluation of the TZ or EPN (e.g., [13]). Ettre emphasized that such results are valid only if one is working in a region and under conditions where these values are independent of the carbon number [12].

In general practice, the separation number (Trennzahl) is more widely used than is the effective plate number. This probably relates to the fact that it is usually easier to measure peak widths at half height than at baseline.

If we use two n-paraffin hydrocarbons to establish the separation number, and remember that these are 100 units apart on the Kováts retention index system, we can calculate the separation number necessary to separate two compounds whose retention indices are known:

$$TZ = \frac{100}{\Delta I} - 1 \qquad (6.6)$$

Hence if we wish to separate butyl-2-methylbutanoate ($I_{100°}^{20\,M}$ = 1234) from n-pentanol ($I_{100°}^{20\,M}$ = 1237), we would require a separation number of

$$TZ = \frac{100}{3} - 1 = 32$$

The effect of temperature and partition ratios on TZ values was discussed earlier; the effects of carrier gas flow rate and column length are perhaps less obvious. One complicating factor is that each partition ratio has a different optimum carrier gas velocity and, where we deal with two widely separated compounds, the optimum velocity for one is not the optimum velocity for the other; the flatter the van Deemter curve, the less serious this problem (Chapter 8). The influence of column length on the separation number was discussed by Ettre [12]. Assuming operation at $\bar{u}_{opt}$, doubling the column length will double the number of theoretical plates. Since resolution is a square root function of the column length [Eq. (1.14)], doubling the column length increases resolution by a factor of $2^{1/2}$ or 1.4. Separation numbers are proportional to $R_s - 1$, but if R_s is reasonably large, $R_s \simeq R_s - 1$ and the separation number increases roughly with the square root of the increase in column length.

Several advantages are claimed for the separation number; the more important of these is that there is no need to know the precise moment of sample introduction, making it a more useful method for splitless and on-column injection, and it can be used to estimate efficiencies under conditions of temperature programming. In this latter case, however, it should be borne in mind that the changing temperature affects relative retentions and the Trennzahl as discussed earlier.

6.3 Other Concepts of Column Performance

Among practicing chromatographers, there has long been dissatisfaction with established methods of evaluating column performance. Neither theoretical nor effective theoretical plate measurements really give any indication of the actual separating power of a column, especially under conditions of temperature programming. The separation number gives a more realistic idea of the resolving power, but the values obtained are limited to a particular region of the chromatogram because they vary with the partition ratios of the test compounds. None of the methods discussed to this point give any indication of the time required for an analysis, and this is in many cases critically important.

Pauschmann [14] suggested a new method for the evaluation of column efficiencies, based on the interrelationship of several chromatographic variables. Kaiser [15] proposed a value termed the "separation power" (SP), based on the sum of the separation numbers of the n-paraffin hydrocarbons over the whole analytical scale from $I = 100$ up to the limit of the system, divided by the adjusted retention time of the last peak:

$$SP = \frac{\sum TZ}{t'_{R_{final}}} \qquad (6.7)$$

If t'_R is expressed in minutes, the separation power would denote the number of totally separated peaks per minute of which the column was capable, which brings the time factor into the expression.

Recently Kaiser offered several new proposals on improved methods of determining column efficiency [15, 16]. Citing arguments against both theoretical plates n and effective theoretical plates N as measurements of column capability, he suggested a new value, n_{real}, which he determined as follows. A series of compounds is chromatographed, and the peak widths at half height are accurately measured and plotted as functions of their partition ratios. The method of least squares is used to calculate the best-fit line for these data. The values of the widths at half height for the hypothetical substances having partition ratios of $k = 0$ and $k = 10$ are determined by extrapolation and assigned the symbols b_0 and b_{10}, respectively:

$$n_{\text{real}} = 5.54 \left(\frac{t'_{R(10)}}{b_{10} - b_0} \right)^2 \tag{6.8}$$

Because $t'_{R(10)}$ is the adjusted retention time for a substance with $k = 10$, and $k = t'_R/t_M$, this is equivalent to

$$n_{\text{real}} = 5.54 \left(\frac{10 t_M}{b_{10} - b_0} \right)^2 \tag{6.9}$$

The value equivalent to the separation number, in this regard, is

$$TZ_{10} = \left(\frac{10 t_M}{b_{10} - b_0} \right) - 1 \tag{6.10}$$

Kaiser then extended this concept to describe a value he termed the "Trennleistungszahl" or "separation power number" TZ_t, which introduced the time factor. This concept hinges on the fact that the peak $TZ_{(10)}$ requires $(10 + 1)t_M$ time units from the moment of injection to the end of the peak:

$$TZ_t = \frac{TZ_{10}}{11 t_M} \tag{6.11}$$

TZ_{10} and n_{real} can be related as follows. Equation (6.6) can be written in the form

$$10_{t_M} = n_{\text{real}} \frac{1}{(5.54)^{1/2}} (b_{10} - b_0) \tag{6.12}$$

and Eq. (6.7) as

$$10_{t_M} = (TZ_{10} + 1)(b_{10} + b_0) \tag{6.13}$$

Equating these expressions and rearranging, we obtain

$$TZ_{10} = 0.425 \left(\frac{b_{10} - b_0}{b_{10} + b_0} \right) (n_{\text{real}})^{1/2} - 1 \tag{6.14}$$

Kaiser's concept can also be used to specify how many evenly spaced peaks within a retention index range of 100 can be separated by a given column, taking into account the polarity of the system:

$$TZ_{100} = \frac{100}{I_2 - I_1} - 1 \qquad (6.15)$$

in which I_2 and I_1 are the retention indices of two separated substances.

Kaiser also proposed a method that he argued presented a more logical basis for evaluating the "true separation quality" of a chromatographic system. Two factors contribute to this, as shown in Figure 6.5. One is the degree of band broadening experienced with increasing partition ratios and is reflected in the slope of the calculated line,

$$\frac{b_{10} - b_0}{10_{t_M}} \qquad \text{(Figure 6.5)}$$

The other is the y intercept b_0, which to a large degree would be influenced by system imperfections such as too low an inlet split

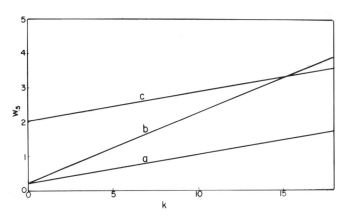

FIGURE 6.5 Graphs of column performance based on Kaiser's proposal [14]. Slopes are governed largely by column efficiency, whereas the intercept is mainly a function of the system. Shown are (a) a highly efficient column in a good system, (b) a less efficient column in a good system, and (c) a highly efficient column in a poor system. Because k is independent of and $w_{0.5}$ dependent on column length, the simple graphs cannot be intercompared for different columns. There have been reasoned arguments against the validity of this concept [17-19] (see text).

ratio, dead volume in the system, poor geometry in the position-ing of the column end with respect to the make-up gas, excessive volume in transfer lines, and similar design defects. Superior results require a column with a low slope in a system with a low intercept.

Franken [17] and Guichon [18] have both argued that peak widths (or peak widths at half height) are not in fact linear func-tions of their retentions (or partition ratios). Smuts et al. [19, 20] report that, although Kaiser's postulate holds true under some experimental conditions, it is not generally valid. They conclude, however, that it should be possible to modify the approach and, by taking the nonlinearity into account, develop a practical pro-cedure for column evaluation.

With these cautionary notes in mind, the method does have some utility as a periodic column check that can alert the inves-tigator to column deterioration or changes in the system such as a broken splitter insert, evidenced by an increased slope or in-creased intercept, respectively. Its utility for comparing columns is limited, because the slope obtained varies with the length of the column used. Assuming a constant phase ratio and the same chromatographic conditions, k is independent of column length, but the peak width is proportional to the square root of the column length. The values given in Table 6.2 were obtained ex-perimentally with sections cut from the same 80-m column. Al-though the values are not in precise agreement with the theoretical relationship, the discrepancy is within the range of experimental error and helps clarify our point. Although the slope m and the b_{10} values for the short column segments are impressive, obviously the intact column is capable of a much higher degree of separa-tion.

TABLE 6.2

L (m)	n^a	h	m	b_0	b_{10}
85	224,300	0.38	0.53	0.58	5.9
10	28,000	0.36	0.19	0.29	2.1
3.5	10,700	0.33	0.15	0.21	1.7

^{a} k of test substance = 5.5.

References

1. Gaspar, G., Arpino, P., and Guichon, G., *J. Chromatogr. Sci.* **15**, 256 (1977).
2. Miller, R. J., and Jennings, W. G., *HRC&CC* **2**, 72 (1979).
3. Golay, M. J. E., *in* "Gas Chromatography 1958" (D. H. Desty, ed.), p. 36. Butterworth, London, 1958.
4. Giddings, J. C., *Anal. Chem.* **36**, 741 (1964).
5. Grob, K., and Grob, G., *J. Chromatogr. Sci.* **7**, 515 (1969).
6. Kaiser, R., *Z. Anal. Chem.* **189**, 1 (1961).
7. Jennings, W. G., and Yabumoto, K., to be published.
8. Rooney, T. A., *Ind. Res. Dev.* **20**(10), 143 (1978).
9. Hawkes, S. J., *J. Chromatogr. Sci.* **7**, 526 (1969).
10. Scott, R. P. W., *in* "Gas Chromatography" (R. P. W. Scott, ed.), p. 144. Butterworth, London, 1960.
11. Hurrell, R. A., and Perry, S. G., *Nature (London)* **196**, 571 (1962).
12. Ettre, L. S., *Chromatographia* **8**, 291 (1975).
13. Grob, K., *Chromatographia* **7**, 94 (1974).
14. Pauschmann, H., *Chromatographia* **9**, 517 (1976).
15. Kaiser, R., *Chromatographia* **8**, 491 (1975).
16. Kaiser, R. E., *Chromatographia* **9**, 337 (1976).
17. Franken, J. J., *Chromatographia* **9**, 643 (1976).
18. Guichon, G., *Chromatographia* **11**, 249 (1978).
19. Smuts, T. W., Buys, T. S., de Clerk, K., and du Toit, T. G., *HRC&CC* **1**, 41 (1978).
20. Smuts, T. W., Buys, T. S., de Clerk, K., and du Toit, T. G., *HRC&CC* **2**, 456 (1979).

CHAPTER 7

TREATMENT OF RETENTION DATA

7.1 General Considerations

Retention behavior is most accurately described as a specific retention volume (Appendix I). This is the volume of carrier gas, reduced to 0°C, required to conduct a solute through the column, and its calculation takes into account the temperature of the column, the flow rate of carrier gas, the pressure drop through the column, and the amount of liquid phase present. Because some of these values are not always known and others (such as the amount of liquid phase) may change with use, specific retention volumes are awkward terms with which to work. Most practicing chromatographers have preferred to deal with relative retentions and to express the retention of a given solute as relative to that of some standard.

It is widely recognized that under isothermal conditions a plot of the logarithm of the adjusted retention time t_R' versus the molecular weight produces a straight line for the members of a homologous series. This fact has been widely used and forms the basis of the retention index systems to be discussed.

7.2 Calculation of the Hold-Up Volume, t_M

With some methods of sample introduction (e.g., splitless, on-column), the precise moment of sample injection is unknown and retention times (or adjusted retention times) cannot be determined by direct measurement. Several methods for calculating the gas hold-up time have been proposed (e.g., [1-3]), and the field has been recently reviewed [4].

In the widely used Peterson and Hirsch approach [5] three members of a homologous series, X, Y, and Z, where the incremental difference in chain length between Y and X is the same as that between Z and Y, are injected as a mixture. The gas hold-up time t_M can be calculated from the relationship

$$t_M = \frac{(t_{R(Y)})^2 - (t_{R(X)} t_{R(Z)})}{2 t_{R(Y)} - (t_{R(X)} + t_{R(Z)})} \tag{7.1}$$

Haken *et al.* [6] emphasized that minor errors in t_M can have a marked effect on retention indices, and compared a number of methods. They reported that a linear least squares analysis as applied to a hydrocarbon series was as accurate as iterative techniques. In our hands, use of the leading edge of a methane peak, as suggested by Rijks [7], has proved most satisfactory. This is in agreement with conclusions reached by Sharples and Vernon [8]. The methane source is discussed in Section 14.2.

7.3   The Kováts Retention Index System, *I*

The retention index system proposed by Kováts [9] has been more widely accepted than any other system; it describes the retention behavior of a compound as equivalent to that of a hypothetical *n*-paraffin hydrocarbon, usually containing a mixed number of carbon atoms. By definition (Appendix I), the index I_A of substance A is given by

$$I_A = 100N + 100n \, \frac{\log V'_{R(A)} - \log V'_{R(N)}}{\log V'_{R(N+n)} - \log V'_{R(N)}} \tag{7.2}$$

where $V'_{R(N+n)}$ and $V'_{R(N)}$ are the adjusted retention volumes of *n*-paraffin hydrocarbons of carbon number N and $N + n$ that are respectively smaller and larger than $V'_{R(A)}$, the adjusted retention volume of A.

In practice, because it is difficult to work with retention volumes and because it is also important to specify the liquid phase used and the temperature of the determination, most workers use retention time measurements and report the index as

$$I_b^a = 100N + 100n \frac{\log t'_{R(A)} - \log t'_{R(N)}}{\log t'_{R(N+n)} - \log t'_{R(N)}} \tag{7.3}$$

where *I* is the retention index on liquid phase *a* at temperature *b*, and the adjusted retention times are quantities analagous to the adjusted retention volumes described for (7.2). It is usually preferable to select the hydrocarbon standards so that $n = 1$. By definition, the retention indices of the *n*-paraffin hydrocarbons are 100N regardless of the liquid phase; for hexane, $I = 600$, and for decane, $I = 1000$. Obviously, graphical means can also be used to establish retention indices. Figure 7.1 shows such a determination. An excellent overview of the Kováts system was published recently [10].

7.4   Effects of Temperature on *I*

Except for large molecular weight compounds at high temperatures, the effect of temperature is not too great for most functional

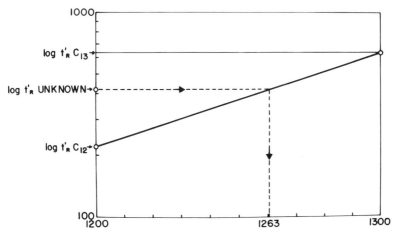

FIGURE 7.1 Determination of the retention index by graphical means using a semi-log plot. The logarithms of the adjusted retention times of the bracketing hydrocarbons are plotted (ordinate) as functions of the retention indices (abscissa), and a line is constructed through those two points. Where the logarithm of the adjusted retention time of the unknown compound intersects that line, a perpendicular is dropped to the abscissa that determines the retention index of the unknown compound.

groups on most liquid phases. In other words, the partition coefficients K_D of most compounds are affected similarly by temperature. The effect is particularly small when the test and reference compounds possess similar functional groups. For compounds of moderate molecular weights, the largest differences in temperature shifts are usually observed between highly polar compounds (e.g., alcohols) and hydrocarbons. Changes in elution order can occur when mixtures containing a variety of functional groups are chromatographed on a polar liquid phase at different temperatures [11] (Figure 7.2). In examining the effect of temperature on the Kováts retention system, Ettre [12] pointed out that the retention indices exhibit a hyperbolic shift as the temperature of the determination is increased. Within a range of perhaps 50°C, however, this shift is approximately linear, and can be expressed as

$$\frac{\Delta I}{10°} = \frac{I_{T_2} - I_{T_1}}{(T_2 - T_1)/10} \tag{7.4}$$

where $\Delta I/10°$ is the incremental change in retention index per 10°C, and T_1 is the lower and T_2 the higher temperature.

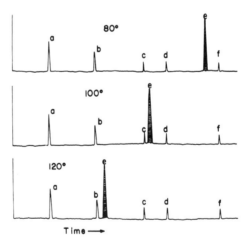

FIGURE 7.2 Effect of separation temperature on relative retentions and elution order: (a) pentyl acetate, (b) isobutyl 2-methyl butanoate, (c) methyl hexanoate, (d) 3,2-methyl butyl propionate, (e) 2-methyl butanol, (f) butyl butyrate. Column, Carbowax 20 M. To facilitate comparison, the 100° and 120°C chromatograms have been expanded horizontally. (From Yabumoto *et al.* [11].) See also Section 8.8.

Larger compounds such as sterols, on the other hand, may exhibit massive shifts in the retention index as the temperature of determination is changed. Values of $\Delta I/10°$ in excess of 30 may be experienced. Möller [13], working on larger-molecular-weight substances including drugs and biologically derived compounds, reported that the retention index was not temperature independent. Interlaboratory deviations were much less if the same temperatures were used for the determination. He proposed standardizing the retention index determination temperature for drugs and unknown substances in biological specimens on OV 1 and OV 17 liquid phases. Lee *et al.* [14] reported that a retention index system based on closely related compounds as internal standards was more reliable for polycyclic aromatic hydrocarbon analysis.

7.5 Other Retention Systems

Göbler [15] pointed out that the Kováts retention index system was not always satisfactory when applied to polar compounds on polar liquid phases, because it required the use of inordinately large hydrocarbons. Because of this problem, several other methods of describing retention behavior have been suggested. Miwa *et al.* [16] and Woodford and van Gent [17] proposed the use of the methyl esters of the straight-chain saturated fatty acids as reference compounds. Results in this system, which is entirely analogous to the Kováts retention index system except for the choice of reference standards, are expressed as the equivalent chain length (ECL) of the hypothetical fatty acid methyl ester that would exhibit that same retention. A carbon-number system based on retentions relative to those of the methyl ketones has been suggested [18], as has one based on normal ethyl esters [19]. The use of similar systems utilizing other references compounds has also been suggested [14, 20].

7.6 Retention Data as Identification Criteria

Since the earliest days of gas chromatography, retention data have been used as criteria of identification. Indeed, the original paper on gas chromatography by James and Martin [21] proposed this application, but because it was difficult to eliminate the

possibility of two compounds exhibiting identical retention behavior in a given gas chromatographic column, the reliability of
such identifications was usually low and they were not well regarded by the discriminating investigator. The practice of "spiking" a sample, or adding a small amount of a known compound
to the sample mixture, with a view to determining whether the
response of the detector to a particular peak is then increased, is
merely the application of retention behavior to component "identification." The high degree of uncertainty associated with reten-
tion-based identifications resulted largely from the limited efficiency of the gas chromatographic column, but batch-to-batch
variations between liquid phases, reactivity of the solid support,
and limited control of column temperature and carrier gas flow
rate all complicated the problem.

The WCOT glass capillary column has lent new significance to
gas chromatographic retention data. Rijks [7] made an intensive
study of the parameters that influence retention indices and reported that with high-resolution glass capillary systems and precise temperature and flow control, indices are highly reproducible. Even between laboratories, errors can be held to less than
0.02%, according to this author.

A considerable number of retention indices are available from
several sources, but most of these data were determined on packed
columns in instruments whose oven temperatures were controlled
with less precision than is possible today. Consequently, the
values obtained from a glass WCOT column in a more precisely
controlled chromatograph may vary from these published data.
Several of the larger industrial laboratories have accumulated impressive files of retention indices, but these are generally regarded
as highly privileged and confidential information. Reliable data
are now beginning to appear in the literature, but to a large
degree this is still scattered and sparse. Rijks reported retention
indices for a number of isomeric low molecular weight hydrocarbons [7], and the indices for 21 aliphatic and 14 heterocyclic
sulfur-containing compounds on packed glass columns containing
Apiezon M, Triton X 305, and polyethylene glycol 1000 at 130°C
have been listed [22]. Peetre and Smith reported values for organ-
ometallics [23] and tetraalkylsilanes [24], and indices for the pyrolysis products from alkyl benzenes on SE 30 WCOT glass columns
are available [25]. Retention indices for a number of polar com-

pounds from flavor isolates, determined on glass WCOT columns containing Carbowax 20 M and SE 30, have been reported [11]. Novotny and Zlatkis [26] reported retention characteristics (in methylene units) for a number of steroidal trimethylsilyl and methoximetrimethylsilyl derivatives on glass capillary columns coated with OV 101 and OV 17. An offering now in press lists programmed temperature retention indices on glass capillary columns (as well as mass spectra) on some 700 flavor and fragrance volatiles [27].

The possibility of two compounds exhibiting identical retention behavior on a given column still exists, although the probability of encountering this phenomenon decreases as the column efficiency increases. By determining retentions on two high-resolution columns of different polarity, the level of confidence in retentions as criteria of identification can be greatly increased. Rijks [7] reported that when the retention indices of isomeric hydrocarbons are plotted in this manner, well-defined groupings result. Figure 7.3 shows a two-dimensional plot of isomeric esters that exhibit this same tendency. Since the conventional electron impact mass spectra of esters within this range of molecular weights (130–200) give at best a barely detectable molecular ion (especially with quadrapole instruments), plots of this type can be very useful in molecular-weight determinations. It is also possible to gain some insight on an unknown compound from a given mixture by its position relative to other groups of compounds. Butanol and isobutanol, for example, have small retention indices on SE 30, but relatively large retention indices on Carbowax 20 M, and their plots are far removed from the esters on the graph. Additionally, the various homologous series exhibit essentially straight-line relationships on the plot, consistent with the fact that the members of a homologous series show a linear relationship between their molecular weights and the logarithms of their retention times. It is also interesting that within these homologous series, the increment of the logarithms for a methylene group is almost equal to that for the n-alkanes (i.e., 100 index units). Several investigators have utilized graphs of this type [28–30].

For the greatest precision, retention indices should be determined isothermally under conditions of close temperature control [7, 31]. With mixtures that embrace a wide range of boiling points, several isothermal runs may be required. Alternatively, linearly

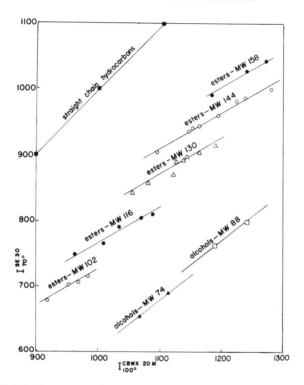

FIGURE 7.3   Retention index on Carbowax 20 M versus the retention index on SE 30 for several compounds. Note the grouping of compounds relative to functional group and molecular weight and the similarity of slope exhibited by groups with similar functional groups. (Adapted from Yabumoto et al.[11].)

programmed temperature runs have been used for such systems (e.g., [14, 27, 32-35]). Provided the program rate is low, such systems seem to work reasonably well although their degree of precision is much less. Said and Hussein [36] described a system for what they term an "absolute retention index" applicable to isothermal or programmed temperature conditions.

References

1. **Ebel, S.,** and **Kaiser, R. E.,** *Chromatographia* **7,** 696 (1974).
2. **Guardino, X.,** and **Albaiges, J.,** *J. Chromatogr.* **118,** 13 (1976).

3. Dominguez, J. A. G., Munoz, J. G., Sanchez, E. F., and Molera, M. J., J. Chromatogr. Sci. **15**, 520 (1977).
4. Kaiser, R. E., HRC&CC **1**, 115 (1978).
5. Peterson, M. L., and Hirsch, J., J. Lipid Res. **1**, 132 (1959).
6. Haken, J. K., Wainwright, M. S., and Smith, R. J., J. Chromatogr. **133**, 1 (1977).
7. Rijks, J. A., "Characterization of Hydrocarbons by Gas Chromatography. Means of Improving Accuracy," Doctoral Thesis, Tech. Univ., Eindhoven, Netherlands, 1973.
8. Sharples, W. E., and Vernon, F., J. Chromatogr. **161**, 83 (1978).
9. Kováts, E., Adv. Chromatogr. **1**, 229 (1965).
10. Ettre, L. S., Chromatographia **6**, 489 (1973); **7**, 39 (1974).
11. Yabumoto, K., Jennings, W. G., and Yamaguchi, M., Anal. Biochem. **78**, 244 (1977).
12. Ettre, L. S., Anal. Chem. **36** (8), 31A (1964).
13. Möller, M. R., Chromatographia **9**, 311 (1976).
14. Lee, M. L., Vassilaros, D. L., White, C. M., and Novotny, M., Anal. Chem. **51**, 768 (1979).
15. Göbler, A., J. Chromatogr. Sci., **10**, 128 (1972).
16. Miwa, T. K., Micolajczak, K. L., Earle, F. R., and Wolff, I. A., Anal. Chem. **32**, 1739 (1960).
17. Woodfard, E. P., and van Gent, C. M., J. Lipid Res. **1**, 88 (1960).
18. Ackman, R. G., J. Chromatogr. Sci. **10**, 535 (1972).
19. van den Dool, H., and Kratz, P. D., J. Chromatogr. **11**, 463 (1963).
20. Novák, J., and Ruzickova, J., J. Chromatogr. **91**, 79 (1974).
21. James, A. T., and Martin, A. J. P., Biochem. J. **50**, 679 (1952).
22. Golornya, R. V., and Garbuzov, V. G., Chromatographia **8**, 265 (1975).
23. Peetre, I. B., and Smith, B. E. F., J. Chromatogr. **89**, 311 (1974).
24. Peetre, I. B., and Smith, B. E. F., J. Chromatogr. **90**, 41 (1974).
25. Svob, V., and Deur-Siftar, D., J. Chromatogr. **91**, 677 (1974).
26. Novotny, M., and Zlatkis, A., J. Chromatogr. Sci. **8**, 346 (1970).
27. Jennings, W., and Shibamoto, T., "Analysis of Flavor and Fragrance Volatiles by Glass Capillary Gas Chromatography and GC/MS." Academic Press, New York, 1980.
28. Tourres, D. A., J. Chromatogr. **30**, 357 (1967).
29. Walraven, J. J., Doctoral Thesis, Tech. Univ., Eindhoven, Netherlands, 1968.
30. Kaiser, R. E., J. Chromatogr. Sci. **12**, 36 (1974).
31. Goedert, M., and Guichon, G., Anal. Chem. **45**, 1180, 1188 (1973).
32. van den Dool, H., and Fratz, P. D., J. Chromatogr. **11**, 463 (1963).
33. Watts, R. B., and Kekwick, R. G. O., J. Chromatogr. **88**, 165 (1974).
34. Majlát, P., Erdos, Z., and Takács, J., J. Chromatogr. **91**, 89 (1974).
35. Halang, W. A., Langlasi, R., and Kugler, E., Anal. Chem. **50**, 1829 (1978).
36. Said, A. S., and Hussein, F. H., HRC&CC **1**, 257 (1978).

CHAPTER 8

TEMPERATURE PROGRAMMING AND CARRIER FLOW CONSIDERATIONS

8.1 General Considerations

The operating temperature of the column, the choice of carrier gas, and the carrier gas velocity are interrelated variables that can exercise profound effects on separation efficiencies and analysis times. Some degree of compromise is usually necessary, first because the optimum values of column temperature and carrier gas velocity are different for different portions of the chromatogram, i.e., they vary with k, and second because columns operated under optimum conditions of temperature and carrier gas velocity may require inordinately long analysis times. Considerable experimentation may be necessary to select a suitable set of conditions.

8.2 Temperature Programming; General Comments

Constituents of a limited boiling point range can usually be investigated under isothermal operating conditions where the highest separation efficiencies are normally realized, with some exceptions (*vide infra*). When a wider range of volatilities is encountered, it is usually necessary to resort to temperature programming. At temperatures sufficiently low to achieve adequate separation of the lower-boiling compounds, the vapor pressures of the higher-boiling components are very low; hence their distribution constants K_D and partition ratios k are excessively large. Not only does this result in long retentions, but because solutes with large distribution constants are associated mainly with the liquid phase [Eq. (1.1)], their concentrations in the moving gas phase are low; hence they must be delivered to the detector more slowly, at lower concentrations that extend over longer periods of time. This results in low, broad peaks, with an adverse effect on sensitivity. By operating the column at a higher isothermal temperature, the vapor pressures of all solutes are increased; the distribution constants and partition ratios become smaller. Because the gas phase concentration of each solute is now higher, it is delivered to the detector at higher concentration, hence more rapidly, and it persists for a shorter time. In this case the peaks are sharp and narrow, and the sensitivity is enhanced accordingly. At this higher temperature, however, the peaks are less widely spaced, and lower-boiling solutes may not have separated because of the limited time they spent in the liquid phase.

Temperature programming combines the benefits of the improved separations achieved at lower column temperatures with the advantage that the detector receives each solute as a sharp, narrow burst. At the beginning of the program, the column temperature is low; lower-boiling components separate, intermediate boiling components move very slowly, and higher-boiling components probably remain cold trapped at the head of the column.

As the column temperature increases, the lower-boiling components have already traversed most of the column under conditions where (presumably) they were effectively separated, and they are now delivered to the detector as sharp, well-defined peaks. As the column reaches these higher temperatures, the intermediate-boiling components begin the chromatographic process and proceed toward the detector while the column temperature continues to increase. As the temperature increases still further, the higher-boiling components begin the partitioning process. Temperature programming causes the partition ratio and the distribution constant of each solute to become continuously smaller. By the time a solute has traversed the column and reached the detector, its K_D has, because of the continual temperature increase, been greatly reduced. Hence its concentration in the gas phase has been increased [Eq. (1.1)], and the detector receives a high concentration for a short time (i.e., a sharp, narrow peak). Temperature programming, then, gives better-defined peaks of higher concentration, resulting in improved sensitivity.

8.3 Program Parameters

Program conditions—initial temperature, initial hold, program rate, final temperature, final hold—are usually selected by trial and error. The initial temperature should be sufficiently low to achieve the desired separation of the lower-boiling components in the mixture; it must be in excess of the minimum temperature for that liquid phase or efficiency will suffer seriously. If the sample contains a cluster of low-boiling components, an initial hold at the starting temperature may be advisable to permit their separation before they are forced to the detector. Complex mixtures of closely spaced peaks may require lower program rates; the relative retention of any two components is inversely proportional to the rate of programming (Section 6.2). The final temperature is selected to achieve sharp peaks from the higher-boiling components, with attention to the maximum temperature recommended for that column. If the maximum desired temperature is reached before all components have eluted, the column can be subjected to an isothermal hold at that temperature unitl the chromatogram is completed.

Some samples for analysis consist of several groups, or clusters, of closely related compound; essential oils serve as an example. Many of these produce chromatograms with an abundance of monoterpene hydrocarbons of similar retention characteristics, a blank area, a cluster of sesquiterpene hydrocarbons, another blank area, and a dispersion of oxygenated terpenes. Multiramp programming to reduce the extent of the blank areas between these clusters is desirable in some such cases.

8.4   Component Separation and Analysis Time

As mentioned above, the degree of component separation decreases as the program rate is increased. Components are also exposed to higher temperatures, but because they move through the column more rapidly, they experience those higher temperatures for shorter periods of time. These interrelated phenomena can be used to good advantage when a column has more efficiency than is required for the separation of a relatively simple system. Under normal operating conditions this results in far more resolution than is required, i.e., widely dispersed peaks separated by expanses of blank chart. There is no great advantage to this over-kill, and the program rate can be increased to a point just consistent with baseline separation with what usually amounts to a significant decrease in analysis time.

8.5   Program Effects on Carrier Gas Velocity

Packed columns utilize relatively large volumes of carrier gas and can be provided with flow regulators that maintain a constant $\bar{u}$ during temperature programming. Open tubular columns operate at much lower flow rates, and gas discharge through an inlet splitter may further complicate the problem of flow regulation. Consequently, open tubular columns are generally operated at constant pressure drop, and the average linear velocity of the carrier gas, $\bar{u}$, changes with column temperature. Figure 8.1 shows the effect of increasing temperature on the average linear carrier gas velocity; in general, as the temperature increases, gas velocity decreases. Hence there is a decrease in velocity during the course of temperature programming in systems operated at constant pressure drop.

It is readily apparent from Figure 8.1 that the decrease in carrier gas velocity per unit change in temperature is more severe with hydrogen than with helium, and more drastic for short columns than for long columns. However, hydrogen produces a much flatter van Deemter curve than does helium (Figure 8.3); similarly, van Deemter curves for short columns are flatter than are those for long columns (see Figure 9.4, Section 9.4). Because of this, decreases in gas velocity are less significant with hydrogen carrier

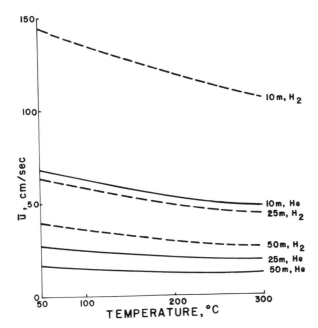

FIGURE 8.1 Effect of temperature on the average linear carrier gas velocity. Columns 0.24-mm i.d. All runs at constant 1.0-atm pressure drop; helium shown as solid lines; hydrogen shown as dashed line.

and/or short columns, provided the average linear gas velocity remains above $\bar{u}_{opt}$.

There have been some suggestions for concurrent pressure programming, i.e., increasing the pressure drop through the column to compensate for the drop in carrier gas velocity that is occasioned by the increase in column temperature; such a course of action would be of doubtful value. It has been mentioned previously that van Deemter curves are k-specific; low-k solutes (those that elute first) exhibit their minima at higher optimum linear velocities than do high-k solutes (*vide infra*). In other words, as the column temperature increases, the gas velocity automatically shifts in the proper direction for optimization. The extent of that shift will be governed by the nature of the carrier gas, the temperature change (or program rate), and the pressure drop through the column (length and diameter), but it is at least in the right direction. Pressure programming would deliver higher pressures, resulting in higher (or constant) gas velocities, as the higher-k compounds are experiencing the major portion of their partitioning processes; the resulting shift is therefore in the wrong direction. As we shall see later, however, it is unlikely that these small changes in carrier gas velocity really have much practical effect, again provided that the velocity does not drop below $\bar{u}_{opt}$ during the course of the run.

8.6   Carrier Gas Choice

Referring back to the Golay equation [Eq. (1.19)], the C term, which reflects the resistance to mass transfer, is the sum of the resistance to mass transfer from the gas phase to the liquid phase (C_G) and from the liquid phase to the gas phase (C_L):

$$h = \frac{B}{\bar{u}} + C_G\bar{u} + C_L\bar{u} \qquad (8.1)$$

In packed columns, longitudinal diffusion in the gas phase (the B term) is critically important. (With packed columns, of course, an A term is also present.) Diffusion is inversely proportional to the density of the carrier gas; less diffusion occurs in denser gases, and nitrogen is widely used as carrier gas. The situation is different in open tubular columns, especially if they are operated at velocities somewhat higher than the optimum (Figure 8.2). Under these conditions mass transfer assumes greater importance. At each theoretical plate of the column, an equilibrium is established between solute molecules in the liquid phase and those in the gas phase; this necessitates a uniform gas-phase concentration through that theoretical plate. As that gas moves to the next theoretical plate of the column, mass transfer of the solute molecules from the gas phase to the liquid phase establishes the equilibrium at this new point. The speed with which these mass transfer steps are accomplished is a function of diffusivity in the gas phase. Less dense gases can be operated at higher velocities because less time is required to establish equilibrium at each theoretical plate.

Figure 8.3 compares van Deemter curves for the three common carrier gases on a glass capillary column [1], and Table 8.1 shows

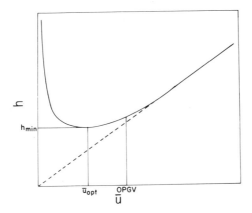

FIGURE 8.2 Typical van Deemter curve showing the optimum flow velocity, $\bar{u}_{opt}$, and the optimum practical gas velocity (OPGV). The latter has been defined [3] as the point where molecular diffusion (that area above the dotted line) amounts to less than 10% of the total value of h.

data gleaned from those curves. Several interesting facts emerge from study of these data.

Operating at the optimum linear carrier gas velocity, efficiencies and analysis times decrease in the order nitrogen > helium > hydrogen. By changing from nitrogen to helium (at $\bar{u}_{opt}$), the

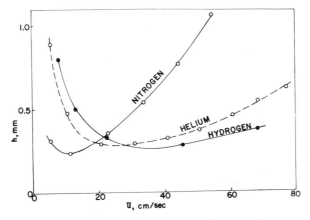

FIGURE 8.3 Van Deemter curves for three different carrier gases; partition ratio of test compound 7.95 (After Rooney [1].)

TABLE 8.1

Effect of Carrier Gas Choice C_{17} at 175°C; $k = 7.95$

$\bar{u}_{opt}$ (cm/sec)	h_{min} (mm)	n_{max} (Σ)	H_{min} (mm)	N_{max}	t_R (sec)
Nitrogen 13	0.22	227,000	0.28	179,000	3446
Helium 21	0.28	178,000	0.36	140,400	2130
Hydrogen 37[a]	0.28	178,000	0.36	140,400	1208

[a]The values shown are from an actual experimental determination. In theory hydrogen would be expected to exhibit slightly larger values of h and H (and correspondingly smaller values for n and N) relative to helium.

system shown loses some 25% of its separation efficiency, but the analysis requires 38% less time. According to these data, hydrogen achieved the same efficiency as helium (in theory it would be expected to be slightly less), but the analysis time is reduced another 43%. Comparing hydrogen with nitrogen, we realize a 65% decrease in analysis time at a cost of 20% of the separation efficiency. In Figure 8.4 [1] these same data are plotted as the number of effective theoretical plates per second as functions of the average linear carrier gas velocity. This method of presentation makes it quite obvious that nitrogen is a poor choice for carrier

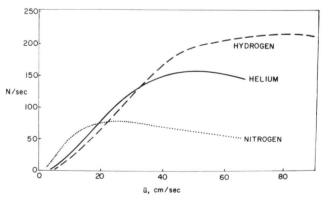

FIGURE 8.4   Data from Figure 8.3 replotted as the number of effective theoretical plates per second as functions of the average linear carrier gas velocities. (After Rooney [1].)

gas in an open tubular column; viewed in this light, hydrogen is the carrier gas of choice, but there are hazards associated with its use (*vide infra*).

Comparison of the curves in Figure 8.3 (and observations of Desty *et al.* [2], among others) indicate that both h_{min} and $\bar{u}_{opt}$ vary inversely with the density of the carrier gas. As discussed above, diffusion occurs more slowly in a dense carrier gas, and a lower average linear velocity allows a longer time for the equilibrium mass transfer to occur at each theoretical plate. Diffusion in the longitudinal direction is of course also slower, and with the substitution of a more dense carrier, the gas velocity can be considerably reduced before band broadening occasioned by longitudinal diffusion becomes limiting. As the carrier gas velocity is increased above $\bar{u}_{opt}$, the increase in h (or loss in column efficiency) per unit increase in $\bar{u}$ is less for the less dense carrier gases; i.e., they exhibit flatter van Deemter curves.

Anything that affects diffusion in the carrier gas—diffusivity of the solute molecules, gas pressure (or average gas density), column temperature—will shift the van Deemter curve and affect both h_{min} and $\bar{u}_{opt}$. In general, the diffusivity of solute molecules varies inversely with their retention times (or partition ratios); i.e., large molecules have low diffusivity and smaller molecules have higher rates of diffusion. Large solute molecules exhibit their $\bar{u}_{opt}$ at lower velocities, where there is more time for their mass transfer, than do small solute molecules, which, because of their higher rates of diffusion, suffer band broadening at this lower velocity. Shorter columns, which for a given velocity have lower pressure drops (the average density of the carrier gas within that column is therefore lower), have flatter van Deemter curves than do longer columns (Figure 8.2). Desty *et al.* [2] investigated columns operated at constant pressure drop but under different system pressures (i.e., from 10 to 9 atm versus from 2 to 1 atm). At the highest system pressure, the value of h_{min} was reduced to approximately 55% of its value at the lowest system pressure, and the van Deemter curve was much sharper.

The effect of temperature is more complicated; viscosity of the gas increases with increasing temperature. Therefore at constant pressure the flow rate decreases as the temperature increases. Again, the effect on separation efficiency is less drastic with short columns and with less dense gases (Figure 8.2).

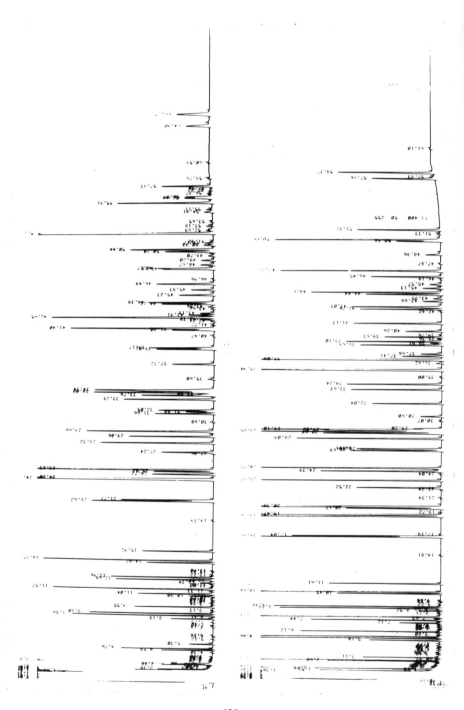

8.7   Optimum Practical Gas Velocity, OPGV

Many investigators tend to become mesmerized by the minimum of the van Deemter curve and point all their efforts toward operation at $\bar{u}_{opt}$. As the carrier gas velocity is increased above this, efficiency drops (i.e., h becomes larger), but this is usually outweighed by the advantages of a shorter analysis time. A flatter van Deemter curve means that less efficiency is sacrificed as the gas velocity is increased (and analysis times are shortened), and hydrogen again emerges as best, with helium a good second (Figure 8.3). This flatter curve can also accommodate a broader range of partition ratios at or near $\bar{u}_{opt}$, as discussed above.

Scott [3] defined an optimum practical gas velocity (OPGV) as the transition point between the curved and linear (high-velocity side) portions of the van Deemter plot (Figure 8.2). For practical purposes, he suggests that this be taken as the point where the longitudinal diffusion term B contributes less than 10% of the total value of h. From this point on, the magnitude of h is largely a function of the C terms. These C values are functions of the square of the column radius, and to a large degree account for the vast superiority of smaller-diameter columns when operated at or above the OPGV. There are reasons to believe that the OPGV is better defined by plots of the type shown in Figure 8.4.

Because mass transfer is also influenced by the temperature, one might predict better column efficiency (smaller h values) at higher temperatures under these flow conditions. Scott [3] argues that for any pair of solutes on any given column an optimum temperature exists at which maximum separation is achieved, but this value can apparently be determined experimentally more accurately than it can be calculated.

In most practical situations the optimum carrier gas velocity is used only to verify the column manufacturer's specifications; analyses are usually better performed at carrier gas velocities of about twice that value, under conditions where efficiency is again traded for speed. Figure 8.5 illustrates the separation of a standard mixture of base-neutral extractable water pollutants on the

FIGURE 8.5   Base-neutral extractables, standard mixture. Column, 15 m × 0.24 mm, SE 54, programmed from 30° to 280°C at 4°C/min. Top, helium carrier, inlet pressure 0.55 kg/cm^2, $\bar{u}_{initial}$ 28.9 cm/sec, $\bar{u}_{final}$ 19 cm/sec. Bottom, hydrogen carrier, 0.55 kg/cm^2; $\bar{u}_{initial}$ 62.5 cm/sec, $\bar{u}_{final}$ 40 cm/sec. (From Jenkins [4].)

same column and under the same program conditions [4]. In one case helium was used as carrier gas, and hydrogen was used in the other. Both gases were supplied at a head pressure of 0.55 kg/cm^2; with helium this attained an average linear velocity of 29.6 cm/sec at T_1 (30°C), dropping to 19 cm/sec ($\bar{u}_{opt}$) at T_2 (280°C). With hydrogen the initial velocity was 62.5 cm/sec, dropping to 40 cm/sec at T_2.

Hence the carrier gas velocity for hydrogen was approximately twice that for helium. Under isothermal conditions this should result in reducing all retention times by a factor of 2 because the partition ratio [Eq. (1.7)] is, for all practical purposes, unaffected by the nature of the carrier gas. With a programmed run, however, the effect is much less dramatic, because higher-boiling components remain cold trapped on the front of the column for a major portion of the analysis time. This means that later-emerging (i.e., high-k) components experience the benefits of the higher carrier gas flow over a smaller fraction of their total residence time.

As discussed in Section 8.6, the rate of diffusion of a given solute through the commonly used carrier gases—and therefore the speed with which an equilibrium distribution is achieved—varies in the order hydrogen > helium ≫ nitrogen (at a given temperature and pressure). The viscosities of these gases lie in the order helium > nitrogen ≫ hydrogen. Viscosity can be regarded as a resistance to flow; when a given pressure drop is applied to a given column at some specific temperature, the average linear carrier gas velocities (i.e., the flow rates) are helium < nitrogen ≪ hydrogen. This is consistent with the data in the legend of Figure 8.5, where at 30°C and a head pressure of 0.55 kg/cm^2 (0.53-atm pressure drop), the average linear carrier gas velocity was 28.9 cm/sec for helium and 62.5 cm/sec for hydrogen. Hence these much higher velocities consistent with hydrogen are obtained at about the same head pressures that are required to deliver the lower velocities used with helium.

8.8   Interrelationships of Column Temperature and Carrier Gas Velocity

The retention temperature of a compound—i.e., that temperature coincident with the peak maximum under a particular set of conditions—is also influenced by both the program rate and the carrier gas velocity. Habgood and Harris [5] found that, by plotting the rate of temperature programming R_p divided by the carrier gas flow rate F as a function of the retention temperature T_E, curves were produced whose shapes were governed largely by the retention characteristics of the compounds. This work was extended to glass WCOT columns [6] to produce curves of the type shown in Figure 8.6. The n-paraffin hydrocarbons were used to produce these data, so that from a knowledge of the Kováts index, one could readily predict the retention temperature and analysis time for a given compound on this column under these conditions.

For the system shown in Figure 8.6, for example, assume a compound whose retention index $I = 1250$, an initial temperature of 100°C, a program rate of 2°C/min, and an average linear gas velocity of 30 cm/sec. Under these conditions that compound will exhibit an elution temperature of 135°C ($2/30 = 0.07$ on ordinate, lying midway between the C_{12} and C_{13} curves), and its retention time will be ($135° - 100°$)/($2°$/min) = 17.5 min. At the same carrier

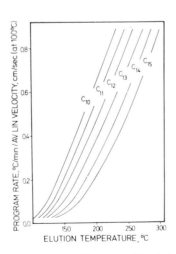

FIGURE 8.6 Plots of program rate/$\bar{u}$ versus the elution temperature of the normal paraffin hydrocarbons. For a compound of known retention index, the elution temperature and analysis time at given program and flow conditions can be predicted from plots of this type [6].

gas velocity, and a program rate of 8°C/min, its elution temperature becomes 175°C and its retention time is 9.4 min.

The interrelationships of temperature programming, carrier gas velocity, and column length can also cause confusion. If a complex mixture containing a range of different functional groups is analyzed on different-length columns but at the same isothermal temperature, the chromatograms are comparable; the resolution of peaks will be different, but the elution order of those peaks will be the same. Changes in the carrier gas velocity will again effect the resolution, but not the elution order.

Under programmed temperature conditions the temperature is changed during the course of the run, and this causes a continuous change in the distribution constant of each solute. As the temperature increases, the retention time of a solute that has a positive retention index increment per 10°C ($\Delta I/10°$) [7, 8] (Chapter 7) decreases at a slower rate than does the retention time of an n-paraffin hydrocarbon. At the same time, the retention time of a solute with a negative $\Delta I/10°$ decreases at a faster rate than does that of the n-paraffin hydrocarbon. All three compounds may exhibit a cross over (change their order of sequence) as the temperature increases while they traverse the column.

To help clarify this point, consider three compounds A, B, and C. Assume that their isothermal elution order (on that same liquid phase, at some temperature T_1) is A, followed by B, followed by C. Let us also assume that their $\Delta I/10°$ values are $A > B > C$. When these same compounds are subjected to temperature programming beginning at T_1, their initial order in the column is A leading, B intermediate, and C last. As the temperature increases, the distribution constants of all three compounds become smaller (concentrations in the moving gas phase increase), and their rates of movement through the column are continuously accelerated. C continues to accelerate at a faster rate than does B, and B continues to accelerate at a faster rate than does A. Hence if they are allowed to spend enough time in the column, B will eventually overtake and pass A, and C will eventually overtake and pass both A and B. Depending on the relative $\Delta I/10°$ values and the time in the column, the elution order may remain A, B, C (short analysis, i.e., short column or high $\bar{u}$), all three compounds may coelute, the order may change all the way to C, B, A (long analysis, i.e., longer column or lower $\bar{u}$), or to any intermediate combination. The time spent in the column will be affected by the rate of temperature programming, the carrier gas velocity, and the column length and phase ratio. Changes in any of these parameters may cause changes in the elution order of any sample not restricted to members of a homologous series. Obviously the dispersion and resolution of those components will also be affected in a manner that can be unpredictable.

Figure 8.7 shows the effect of temperature programming on column separation efficiency as determined by the separation number. Obviously, much higher separation efficiencies are achieved at low program rates. This relates to the fact that relative retentions usually vary inversely with column temperature. Hence as the temperature increases, relative retentions and separation numbers (or Trennzahl) become smaller. These relationships were explored in Section 6.2.

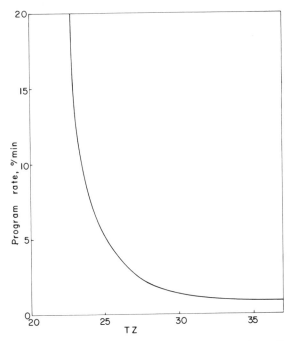

FIGURE 8.7 Effect of temperature programming on column efficiency; 0.25 mm × 40 m. WCOT Carbowax 20 M column. Test compounds, *n*-heptanol and *n*-octanol, 70°–170°C at the indicated program rates.

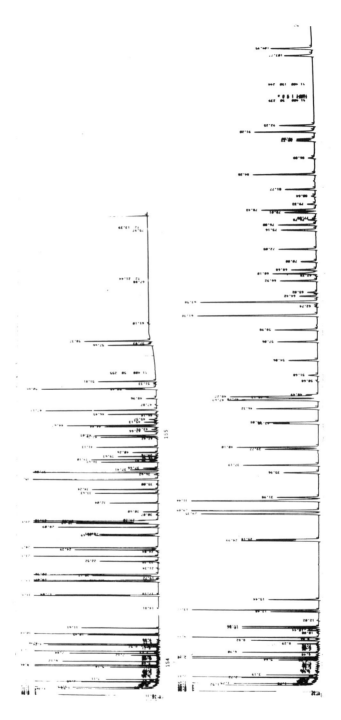

The effect of a higher program rate is usually greater on compounds of larger relative retention; the change in separation efficiency with compounds of shorter retention is of course in this same direction, but the magnitude of that change is usually less. In general, raising the program rate results in a decrease in separation efficiency, shorter analysis times, and higher sensitivity. Shrewd interpretation on the part of the analyst can help exploit these relationships to achieve just that degree of separation necessary. Figure 8.8 shows chromatograms of a standard mixture of base-neutral extractable water pollutants under conditions identical to those used in Figure 8.5, except for changes in the program rate [4].

8.9 Hazards Associated with Hydrogen Carrier Gas

Some of the advantages of hydrogen as a carrier gas include the facts that it is the least k-specific of the carrier gases (Figure 8.3) and that it gives the highest efficiency per unit time (Figure 8.4). In addition, because equivalent efficiencies are obtained at higher average linear carrier gas velocities, compounds exhibit lower elution temperatures (and shorter retentions) during temperature-programmed runs (Section 8.7). It should be noted that in accordance with this and, as discussed earlier, the elution order of dissimilar compounds may well differ from that observed with other gases at lower velocities (all other conditions, i.e., column and program rate, being constant). Hydrogen is also much less expensive than helium, and it is very doubtful whether oxygen scrubbers are required on the carrier gas line when hydrogen is used as carrier (Chapter 14). Occasionally someone questions the possibility of sample components or liquid phase interacting with hydrogen; the prospects of this occurring in the absence of a suitable catalyst are, for all practical purposes, nonexistent [9].

There is an obvious hazard associated with hydrogen—it can form explosive mixtures with air over a fairly wide range of con-

←——————————————————————————

FIGURE 8.8 Base-neutral extractables from water; standard mixture. Column, 15 m × 0.24 mm SE 54, hydrogen carrier at $\bar{u}_{initial} = 62.5$ cm/sec, $\bar{u}_{final} = 40$ cm/sec. Injections, 2 μl splitless, hot needle method. Top, programmed from 30° to 280°C at 4°C/min. Bottom programmed from 30° to 280°C at 2°C/min. (From Jenkins [4].)

centrations. Some workers have argued that because of its high diffusivity, under most well-ventilated conditions, it would be difficult to accumulate enough hydrogen to form an explosive mixture, even within most chromatographic ovens. Others feel that the danger is real and recommend a number of precautionary measures. These include the following.

Hydrogen gas supply: Tanks should be stored outside the laboratory, away from other gas cylinders. The hydrogen supply line should be metal, protected from mechanical injury, and its diameter should not be greater than that necessary to deliver the required gas volume; rupture of a $\frac{1}{16}$-in. line would be less hazardous than rupture of a $\frac{1}{4}$-in. line. Flow monitors can be designed to sound alarms, shut down the carrier, and even divert nitrogen or carbon dioxide into the carrier gas line in the event of a larger-than-preset gas discharge (indicative of a ruptured line or broken column).

Inlet splitters: The splitter outlet should be activated only during injection; some discharge the split stream into the flame of an ignited burner; others vent that stream into a fume hood, ensuring rapid dilution.

Ovens: These are of major concern. If a column should break at the detector end, the column itself offers enough flow restriction that the prospects of discharging hydrogen fast enough to accumulate an explosive mixture are probably slight. If the column should break at the inlet end, gas would be discharged into the oven at a much faster rate. Flow monitors in the supply line (see above) offer protection, and commercial leak detectors can be modified to sample the oven [10] atmosphere continuously and either sound an alarm or shut down the system before an explosive concentration has been achieved. The latter can be designed to disconnect the oven heater circuit, shut down the hydrogen line, and/or divert nitrogen (or carbon dioxide) into the carrier gas lines to achieve a rapid dilution (and protect the column). Some degree of continuous oven venting is desirable, and oven doors should be spring-loaded rather than rigidly latched.

References

1. **Rooney, T. A.,** *Ind. Res. Dev.* **20**(10), 143 (1978).
2. **Desty, D. H., Goldup, A.,** and **Whyman, B. H. F.,** *J. Inst. Pet., London* **45,** 287 (1959).

3. **Scott, R. P. W.**, *in* "Gas Chromatography" (R. P. W. Scott, ed.), p. 144. Butterworth, London. 1960.
4. **Jenkins, R.**, personal communication (1979).
5. **Habgood, H. W.**, and **Harris, W. E.**, *Anal. Chem.* **32**, 450 (1960).
6. **Jennings, W. G.**, and **Adam, S.**, *Anal. Biochem.* **69**, 61 (1975).
7. **Ettre, L. S.**, *Anal. Chem.* **36**(8), 31-A (1964).
8. **Yabumoto, K.**, **Jennings, W. G.**, and **Yamaguchi, M.**, *Anal. Biochem.* **78**, 244 (1977).
9. **Grob, K.**, and **Grob, G.**, *HRC&CC* **2**, 109 (1979).
10. **Olufren, B.**, *HRC&CC* **2**, 578 (1979).

CHAPTER 9

SPECIAL ANALYTICAL METHODS IN GLASS CAPILLARY GAS CHROMATOGRAPHY

9.1 General Considerations

Several different approaches have been used to develop additional information from a gas chromatographic separation. Retention characteristics as determined on two different liquid phases, differential temperature adjustments in two sequential columns containing different liquid phases, the use of subtractive precolumns or selective detectors, and recycle chromatography have all been explored. The use of subtractive precolumns and selective detectors has been covered elsewhere [1]; a few selected examples of developments in the other enumerated areas will be discussed below.

9.2 Two-Dimensional Techniques

As pointed out by Bertsch [2], the term "two-dimensional GC" has been applied to a vast range of systems. According to this

author, the term should be restricted to systems that entail either
(1) "two columns of different selectivity in combination with a
system (integration, MS identification, etc.) which will permit
assignment of retention indexes" or (2) "two columns of different
selectivity and a device (prep-scale collection tube, valve, etc.) to
selectively transfer a portion of a chromatographic run from one
into another column." These definitions would include most applications that have been termed "heart-cutting."

A number of workers have contributed to the development and
exploration of these techniques (e.g., [2–6]). A typical application,
in which one restricted region of a chromatogram is transferred
to a second dissimilar column for further resolution in a second
liquid phase, is illustrated in Figure 9.1 [6]. Figure 9.2 shows a

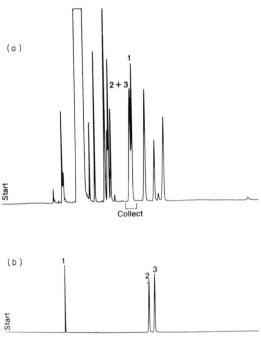

FIGURE 9.1   Two-dimensional GC. (a) Chromatogram A, (b) chromatogram B.
One cluster (top) in an impure fraction of ethyl benzene, collected from a glass
capillary column coated with SP 2100 and (bottom) shunted to a second glass
capillary coated with SP 1000, where separation is achieved. (From Miller *et al.*
[6].)

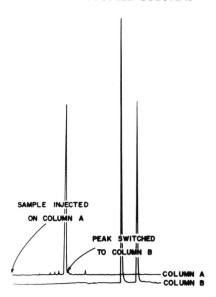

FIGURE 9.2   Two-dimensional GC. Top: chromatogram of a mixture of *n*-pentyl acetate and isobutyl isobutyrate eluting as a single peak from a glass capillary column coated with methyl silicone, (bottom) valve-shunted to a second glass capillary column coated with Carbowax 20 M to achieve separation and precise retention indices on the two dissimilar liquid phases. (From Jennings *et al.* [5].)

peak emerging from a nondestructive detector attached to a methyl silicone column that is then shunted to a second column coated with Carbowax 20 M. Retention indices can thus be determined on a given peak(s) on two different liquid phases.

9.3   Serial Coupled Columns at Different Temperatures— SECAT

If two columns of similar length but containing different liquid phases are serially coupled and mounted so they can be operated at different isothermal temperatures, minor changes in column temperatures can have a striking effect on the dispersion pattern of mixtures containing compounds with different functional groups [7, 8]. Very minor changes in the temperature of one column can cause major changes in the chromatographic pattern, and Kaiser and Rieder [8] recommend temperature control toler-

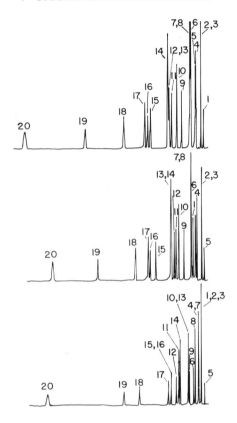

FIGURE 9.3   SECAT chromatograms, with sequential columns. Column A, OV 101; column B, Carbowax high-polymer. (Top) both columns at 70°C; (middle) A at 73.5°, B at 70°C; (bottom) A at 70°, B at 92°C. Chart speeds and other parameters same. (After Kaiser and Rieder [8].)

1.	acetonitrile	11.	octene-1
2.	hexane	12.	octane
3.	isopropylether	13.	n-butanol
4.	tetrahydrofuran	14.	toluene
5.	acetone	15.	methylpentanoate
6.	heptane	16.	nitropropane
7.	n-propanol	17.	cyclopentanone
8.	benzene	18.	nonane
9.	methylbutyrate	19.	methylhexanoate
10.	dioxan	20.	decane

ances of less than 0.2°C. Figure 9.3 illustrates the technique as applied to a model system. Suggestions for the computerized calculation of that system to achieve the best dispersion [8], based on the ABT concept [9] have also been proposed.

9.4   Multiple Pass (Recycle) Chromatography

The linear velocity of the carrier gas through the column is of course not constant. Because of the pressure drop through the column, velocity at the inlet (high compression) end is lower than the average velocity, and the velocity at the outlet (decompression) end is higher than the average velocity. Many workers have concluded that shorter columns owe their greater efficiency per unit length to the fact that they can be operated at or near the optimum linear carrier gas velocity over a greater fraction of that column length; this assumption is of doubtful validity. Giddings [10] argued that if the *average* linear carrier gas velocity is optimized, changes in the diffusion coefficient due to decompression as the gas passes through the column compensate for the negative effects accompanying the decrease in gas velocity; Sternberg [11] supported this reasoning.

Shorter columns do, however, exhibit flatter van Deemter curves (Figure 9.4) [12]. Hence, although their efficiencies per unit

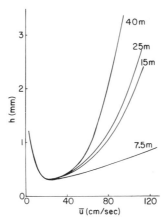

FIGURE 9.4   Van Deemter curves from different length segments of the same OV 101 coated glass capillary column. (After Yabumoto and VandenHeuvel [12].)

length are essentially the same at $\bar{u}_{opt}$ (all h values are essentially equal), shorter columns lose efficiency less rapidly as the carrier gas velocity is increased. (This assumes the columns are identical in every respect except length.) Figures 9.5 and 9.6, reconstructed from data presented by Yabumoto and VandenHeuvel [12], illustrate this point. Shorter columns, then, should exhibit much higher OPGV values [13, 14] (Chapter 8); the one limitation of short columns lies in the fact that they can deliver only a limited number of theoretical plates.

Recycling through a short column should allow the combination of higher OPGV with larger theoretical plate numbers. Such a system designed by Jennings et $al.$ [15, 16] achieved in excess of 2,000,000 theoretical plates in slightly less than 16 min when operated at $\bar{u}_{opt}$ and over 3600 theoretical plates per second at OPGV. As shown schematically in Figure 9.7, the unit offers several additional advantages. Built around a mechanical valve

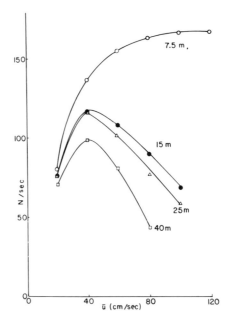

FIGURE 9.5 Plots of the number of effective theoretical plates per second as functions of the average linear carrier gas velocities for different length segments of the same glass capillary column, calculated for C_{14}, $k = 6.5$ at 130°C.

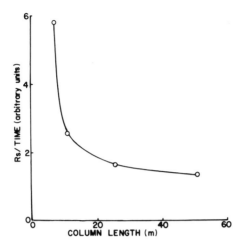

FIGURE 9.6   Resolution per unit time (C_{13}-C_{14}, 140°C) as a function of column length. (Constructed from data shown by Yabumoto and VandenHeuvel [12].)

rather than Dean's switches [17], flow conditions in the recycle apparatus are independent of those in the analytical column, which can therefore be used in a programmed mode. Those areas of the chromatogram that can be sufficiently resolved by the analytical column need not be directed into the recycle unit, a considerable economy of analysis time. Areas that are inadequately separated can be shunted to the unit and recycled; column length (and theoretical plate numbers) are for all practical purposes converted to operational parameters by this apparatus.

Another major advantage can be realized with recycle chromatography. Laub and Purnell [18, 19] described a process they termed "window diagramming" that established what binary mixture of any two liquid phases would achieve the best separation of any given mixture. The prediction required the preparation of several packed columns containing different proportions of those two liquid phases, and the preparation of the analytical column containing the calculated proportions. Jennings *et al.* described an alternative approach [16]. Two recycle units containing dissimilar liquid phases could be interconnected with an analytical column (Figure 9.8). Window diagrams could be constructed from data collected by varying the time spent in liquid phase 1 compared to the time spent in liquid phase 2. From that diagram

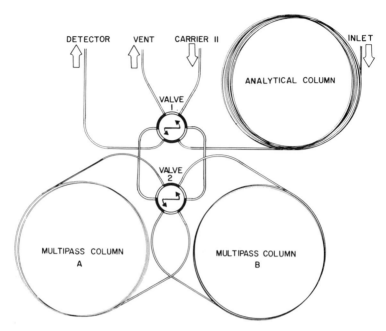

FIGURE 9.7   Schematic of a glass capillary recycle system which has yielded in excess of 2,000,000 theoretical plates. (After Jennings *et al.* [16]).

one could predict what ratio of recycle passes in each of the two liquid phases would be required to deliver the required binary mixture of liquid phases; the total number of passes would be selected to deliver the plate numbers required for the separation. In this case both column length and column polarity have been converted to operational parameters.

9.5   Limitations in Recycle Chromatography

A limited range of partition ratios can be accommodated within the recycle unit. It has been shown [15] that, using two precisely matched column segments in the recycle unit, the number of cycles to which a given range of partition ratios can be subjected is

$$n_{cyc} = \frac{k_B + 1}{k_B - k_A} \tag{9.1}$$

A major problem with the operation of the apparatus is deciding just when the recycle valve should be activated. Any error in the switching interval is cumulative and becomes evident when a portion of a peak is excluded from the mainstream and proceeds to the detector. A nondestructive in-line detector whose flow

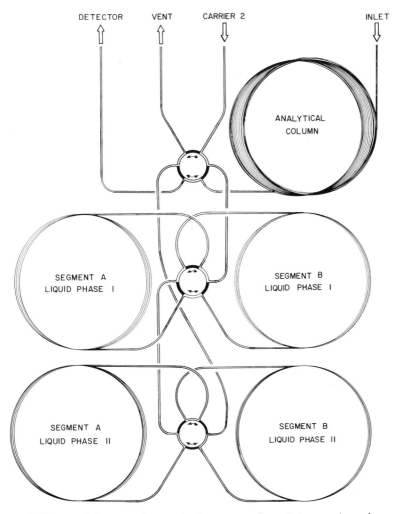

FIGURE 9.8 Schematic of a recycle chromatography unit transposing column polarity to an operational parameter. (After Jennings *et al.* [16].)

characteristics matched those of the column (i.e., which did not contribute band broadening) could be used not only to trigger the switching valve, but also to help in the solution of the third problem.

This third limitation relates to in-column peak broadening. Even with a flow-switching device that does not contribute to band broadening, the width of a peak relative to its height must increase as a function of the column length traversed. In other words, the height of a peak continually diminishes as recycling continues, and its concentration eventually falls below the limits of detection. The apparatus described above [15] was able to achieve some 20 passes on major components before the peak concentration became limiting. It should be possible to focus the band thermally by inserting a short length of coated platinum column into the unit, one end of which could be cooled thermoelectrically to establish a short-lived temperature gradient. As a band entered this zone and proceeded down the temperature gradient (toward the colder zone), the partition ratio of the front end would continually increase more rapidly than the partition ratio of the back end, and the band should be narrowed. Before the sharpened band passed over the coldest point and the band-narrowing process reversed, the zone should be uniformly and instantaneously heated to column temperature (or higher)—again thermoelectrically. Development of a suitable in-line detector would permit the cooling and heating to be delivered as timed pulses, but the constraints placed on this detector are indeed stringent. Thermal focusing would have the effect of decreasing the distance between these partially resolved components—i.e., their apparent relative retentions—but the entire band would be sharpened. Provided that the resolution of the partially resolved components was not adversely affected, thermal focusing should raise their concentrations to a point where they could be further recycled. The feasibility of thermal focusing to produce sharper band of greatly increased solute concentration can then be returned to recycle has been recently demonstrated [20].

References

1. **Jennings, W.,** and **Shibamoto, T.,** "Analysis of Flavor and Fragrance Volatiles with Glass Capillary Columns and GC/MS," Academic Press, New York, 1980. (In press).

2. **Bertsch, W.**, *HRC&CC* **1**, 85, 187, 289 (1978).
3. **Rijks, A. J.**, "Characterization of Hydrocarbons by Gas Chromatography. Means of Improving Accuracy," Doctoral Thesis, Tech. Univ., Eindhoven, Netherlands, 1973.
4. **Schomburg, G., Husmann, H.**, and **Weeke, F.**, *J. Chromatogr.* **112**, 205 (1975).
5. **Jennings, W. G., Wyllie, S. G.**, and **Alves, S.**, *Chromatographia* **10**, 426 (1977).
6. **Miller, R. J., Stearns, S. D.**, and **Freeman, R. R.**, *HRC&CC* **2**, 55 (1979).
7. **Pretorius, V., Smuts, T. W.**, and **Moncrieff, J.**, *HRC&CC* **1**, 200 (1978).
8. **Kaiser, R. E.**, and **Rieder, R. I.**, *HRC&CC* **2**, 416 (1979).
9. **Kaiser, R. E.**, "Chromcard ABT-Konzept." Inst. Chromatogr., P.O. Box 1307, D 6702 Bad Dürkheim, West Germany.
10. **Giddings, J. C.**, *Anal. Chem.* **36**, 741 (1964).
11. **Sternberg, J. C.**, *Anal. Chem.* **36**, 921 (1964).
12. **Yabumoto, K.**, and **VandenHeuvel, W. J. A.**, *J. Chromatogr.* **140**, 197 (1977).
13. **Scott, R. P. W.**, and **Hazeldean, G. S. F.**, *in* "Gas Chromatography 1960" (R. P. W. Scott, ed.), p. 144. Butterworth, London, 1960.
14. **Desty, D. H., Goldup, A.**, and **Swanton, W. T.**, *in* "Gas Chromatography 1962" (N. Brenner, J. E. Callen, and M. D. Weiss, eds.), p. 105. Academic Press, New York, 1962.
15. **Jennings, W., Settlage, J. A.**, and **Miller, R. J.**, *HRC&CC* **2**, 441 (1979).
16. **Jennings, W., Settlage, J. A., J. A., Miller, R. J.**, and **Raabe, O. G.**, *J. Chromatogr.* **186**, 189 (1979).
17. **Deans, D. R.**, *Chromatographia* **1**, 18 (1968).
18. **Laub, R. J.**, and **Purnell, J. H.**, *Anal. Chem.* **48**, 799 (1976).
19. **Laub, R. J.**, and **Purnell, J. H.**, *Anal. Chem.* **48**, 1720 (1976).
20. **Jennings, W. G., Settlage, J. A., Ingrahan, P. I.**, and **Miller, R. J.**, Paper No. 2, Expo Chem 79, Houston, Texas, 22–25 October (1979).

CHAPTER 10

COLUMN STABILITY

10.1 General Considerations

Thizon *et al.* [1] defined the useful life of a packed column as that time after which only half the initial amount of liquid phase remains in the column. They reported that column life was related to operating temperature, parameters such as the concentration of oxygen and water in the carrier gas, and the nature of the solid support. In open tubular columns deterioration may be evidenced as a loss in efficiency (i.e., broad peaks), a decrease in partition ratios (loss of liquid phase), changes in polarity (degradation of liquid phase), tailing peaks, or a noisy or drifting baseline. In a gross sense, the useful life of an open tubular column is related to the condition—chemical and physical—of the liquid phase. As long as that liquid phase is unchanged chemically (which would affect its polarity), remains clean (no residues from injected samples, septum particles, etc.) and in the form of a thin, continuous, and uniform film, the column should continue to perform with good efficiency. At this point we will consider difficulties associated with physical changes in the liquid phase; difficulties caused by other problems such as deposits from injection residues will be considered later.

Good columns are good because they have an absence of "bad

FIGURE 10.1   A good coating is perfectly uniform and even.

spots." This has long been recognized, and Cramers *et al.* [2] recently emphasized that the column efficiency depends less on the phase ratio than on globules or spots in the column that have the largest effective film thickness, localized though they may be. Hence the useful life is related to the degree of attraction between the liquid phase film and the supporting glass surface. As long as the liquid phase endures as a thin uniform film, the column should continue to perform with good efficiency. As the liquid phase retreats from the surface to form discrete globules of liquid phase, column efficiency will drop sharply (Figures 10.1–10.3). Occasionally column deterioration is restricted to one or two coils, must usually at the inlet end of the column. Removing this section of the column will usually restore efficiency, and the loss in column length will rarely be of concern. An understanding of the interrelated factors that bring about these changes can help in prolonging column life.

A great deal of work in the field of detergency has been concerned with the deposition and removal of thin films on hard surfaces, and the energy relationships between the film and the surface have been probed by several investigators. Much of this,

FIGURE 10.2   The beginning of deterioration. Note surface imperfections in the coating.

FIGURE 10.3 A deteriorated column. Note the appearance of distinct globules.

as it relates to WCOT glass technology, has been reviewed and summarized [3].

There are essentially two groups of forces that concern us: one is the forces of attraction that exist between the liquid phase and the glass surface; the other, viewed from the standpoint of the glass surface, we might term the repulsive forces, the most critical of which is probably the cohesive force of the liquid phase itself. By forming globules, the surface area and the free surface energy of the liquid phase can be reduced to a lower value, which is of course a more natural state. As long as the forces of attraction dominate, the liquid phase endures as a thin film. When these forces are no longer superior, the liquid phase retreats from the surface and globules appear. Two factors that are of primary concern in this regard are column temperature and exposure of the surface to materials that would be more strongly adsorbed than is the liquid phase.

The method of column manufacture, i.e., the means used to deposit the coating, seems to play a role in determining the resistance of the coating to higher temperatures. This probably relates to the intimacy of contact achieved in the coating operation and to whether the liquid phase is actually deposited on glass, hydrated glass, or an intermediate layer of salt crystals.

10.2   Carrier Gas Considerations

Columns subjected to heat in the absence of carrier gas flow degrade rapidly; carrier gas should always be turned on before the oven is heated, and the oven should always be cooled before the carrier gas is shut off. Entrained dust (which can originate from molecular sieves or other sources) or any other particulate matter in the carrier gas will also lead to a rapid loss in column efficiency, and in-line dust filters should be employed.

The need for gas driers on the carrier gas line seems to relate to the ability of the water in the gas to contact the glass surface of the column and displace the layer thereon. This is probably a function of both the water permeability of that liquid phase and the coating technique that was used in column preparation. For smooth-bore columns coated by the heated-inlet-tube modification of the Golay technique (Chapter 3), gas driers are rarely required with gum-type phases, such as OV 1, SE 30, SE 52, and SE 54, or silicone fluid phases, such as SF 96. (The latter is marketed in more highly purified form as OV 101 and SP2100.) For columns coated by this method the requirement for a dry carrier gas increases with the polarity of the liquid phase. OV 17 (equivalent to SP 2250) columns exhibit much longer useful lives when used with dry carrier gas; dryness becomes increasingly important to Carbowax 20 M, FFAP, and SP 1000 columns, and it is critical for polyester coatings.

On the other hand, just the opposite situation prevails with columns that have been coated over a barium carbonate deposit; such columns coated with polar liquid phases will reportedly withstand direct aqueous injections, but those containing apolar liquid phases are less tolerant of such treatment [3].

The action of water on smooth-bore columns can be explained by the ability of the water to penetrate the liquid phase, contact the glass surface, and displace liquid phase. Silicones are widely used to impart waterproof characteristics to fabrics and leathers; provided that the silicone liquid phase endures as a thin but continuous film, it is doubtful that water is able to contact the glass surface. As the polarity of the liquid phase increases, so does the permeability to water, which may relate to the increased need for the protection offered by the gas driers. Water has a lower affinity for barium carbonate than it does for glass, and although it can penetrate the polar coatings to contact the underlayment, the liquid phase is apparently more strongly attracted to that deposit than is the water. The behavior of silicone phases on barium carbonate columns is more difficult to explain but probably relates to these same principles.

The interrelationships of gas drier needs and column coating are in all probability also influenced by the continuity of the liquid phase film and the tenacity with which it is held. Sparsely covered sites or loosely deposited films may allow points for the initiation of preferential displacement by water [4]. The lifetime of a gas drier is not infinite, and it should be replaced or regenerated periodically.

Similarly, oxygen scrubbers are important for the protection of those liquid phases that are capable of being oxidized (e.g., Carbowax phases, FFAP, etc.), but they are a needless expense for some others (e.g., SP 2100). Both nitrogen and helium normally contain appreciable concentrations of oxygen; where hydrogen is used as carrier gas, it is highly probable that the carrier is oxygen free, and oxygen scrubbers can usually be omitted.

10.3 Temperature Effects

In most cases column life is inversely proportional to the temperature to which that column is exposed. Higher temperatures are less deleterious if they are maintained for shorter times; hence it is less harmful to program the column to that temperature and cool it rapidly than to subject it to the elevated temperature for a period of time. With silicone-type (and sometimes with other) liquid phases, elevated temperatures may cause a loss of the deactivation chemicals before the liquid phase film itself suffers physical deterioration.

Polydimethylsiloxane liquid phases are available as gums (OV 1, SE 30) or fluids (OV 101, SP 2100, SF 96, DC 200). These usually exhibit an upper temperature limit of about 300°C, although column life will be extended significantly if the upper temperature limit is lowered to about 280°C. In many cases the deactivation properties are lost at temperatures of 220°–250°C. This behavior varies with the deactivation treatment used by the column manufacturer, and when it does occur nonsorptive compounds such as hydrocarbons will continue to produce well-formed peaks, but tailing will be exhibited by more active compounds, such as alcohols and ketones. The deactivation can usually be restored by the method of de Nijs et al. [5] (vide infra). The gum-type phases (e.g., OV 1, SE 30) are usually more stable than are the liquids [6]; higher coating efficiencies are normally achieved with OV 101 or SP 2100. SF 96 and DC 200 may have higher levels of residual catalyst and impurities; they usually require longer curing and are less satisfactory for GC/MS applications.

Polyphenylethylsiloxanes include SE 52 (gum, 5% phenyl), DC

10 (liquid, 35% phenyl), OV 17 or SP 2250 (liquid, 50% phenyl), and SE 54 (gum, 1% vinyl, 5% phenyl). SE 52 and SE 54 are similar to the methyl silicones, in that they exhibit good stability at 280°C and will tolerate higher temperatures for short times; both perform well in GC/MS applications and are reasonably tolerant of overloading and large injections. As the amount of phenyl is increased, some stability is lost; OV 17 (and SP 2250) exhibits upper practical temperature limits of about 250°C.

The polycyanopropylsiloxanes are more polar silicone phases. The OV 225 (liquid, 25% cyanopropyl, 25% phenyl), Silar 10C, and SP 2340 (liquid, 75% cyanopropyl) all have upper practical limits of about 250°C.

Several polyethylene glycol-type liquid phases are mixtures of ethylene oxide condensation units, and are named to reflect the average of a variable range of molecular weights. Hence there can be large batch-to-batch differences. More restricted molecular weight ranges can be prepared by gel permeation chromatography, but columns prepared from these fractions do not seem to exhibit superior characteristics. Carbowax 20 M (waxy solid) is widely used with glass capillaries. Terephthalic acid-terminated Carbowax 20 M has been termed the free fatty acid phase (FFAP); SP 1000 is almost the equivalent. Both exhibit practical upper temperature limits of about 220° and lower limits of about 60°C. Another series of polyethylene glycols, characterized by higher molecular weights and bearing the name "Superox," is also available. Recent reports [7] indicate upper temperature limits of about 260°C for the highest molecular weight fraction, and slightly lower limits for the other members of that series.

All these temperatures should be regarded as approximate rather than absolute values. Some users report exceeding these limits; one applications chemist asserts that in demonstrating GC/MS units he routinely programs SP 2100 columns up to 350°C, but those columns would probably exhibit much longer lives if lower limits were employed. Fused silica columns that exhibit higher temperature limits for all these liquid phases are appearing on the market at this time.

10.4   Effect of Injection Size and Sample Composition

Another—and interrelated—factor influencing column life is the type of materials to which the column is exposed. Large injections may have a scrubbing effect that displaces some liquid phase; strongly adsorbed materials such as water or alcohols will, if they can penetrate the liquid phase and contact the glass surface, be more strongly adsorbed and displace liquid phase. Certain solvents (e.g. carbon disulfide) are very efficient at displacing the liquid phase. Low levels of moisture in the carrier gas will have a deleterious effect on column life, particularly on liquid phases such as Carbowax 20 M. SE 30, on the other hand, is more impervious to water and makes it more difficult for the water to reach the glass surface, where it would adsorb and lessen the attractive forces holding the liquid phase. It is very probable that there is an interrelationship between these two areas. At a higher temperature, the effect of small carbon disulfide injections will be much more devastating than they would be at lower temperatures. The increased temperature stability which has been reported for columns containing Silanox 101, either as a precoating deposition or when admixed with the coating solution, may result from the hydrophobicity of the silicon dioxide as well as changes in contact angle relationships [8].

Both splitless injection, in which a relatively large volume of solvent is deliberately condensed on column (Section 4.3), and on-column injections (Section 4.5) will lead to more rapid column degradation. Solvents used in such injections should be selected with attention to their affinity for adsorption on glass (i.e., the probability of liquid phase displacement) and their solubility in or reactivity with the liquid phase used. Methanol is a very poor choice for a wide range of liquid phases, including Carbowax 20 M and SP 2100, and its use in splitless or on-column injection can lead to rapid column deterioration; acetone tends to degrade silicones and its use should be avoided with those liquid phases. Carbon disulfide, which exhibits a high heat of adsorption with glass, has already been mentioned.

10.5 Chemical Deterioration

Another form of column deterioration involves chemical changes in the liquid phase itself. Sometimes these are caused by reaction with injected materials (e.g., silylation reaction mixtures on susceptible liquid phases); more common are deteriorative changes engendered by oxygen (or air) in the carrier gas. The polyethylene glycol liquid phases—Carbowax, FFAP 9 or SP 1000,

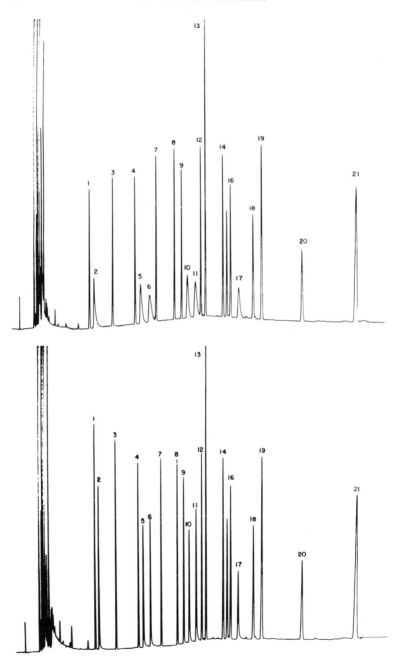

and Superox—are especially susceptible to oxygen. Usually as a liquid phase oxidizes, it becomes more acidic, it may discolor and take on a yellow or brownish tint, and its polarity increases. This latter is not too surprising; chemical changes of almost any type will usually be accompanied by a change in liquid phase polarity. Daily injections of suitable test solutions (Section 5.6) can help guard against this problem. When a sudden polarity shift is observed, it should be possible to correlate this occurrence with a change in the analytical procedures, the installation of a new tank of carrier gas (contaminated gases are no rarity), or an exhausted oxygen scrubber; prompt remedial action may save an expensive column.

10.6 Salvaging Columns

Columns that exhibit good conformation on some peaks and tailing or attrition of others have not been properly deactivated or have lost at least a portion of their deactivation because of use conditions. Figure 10.4 shows an SE 52 column used in the analysis of methyl esters of fatty acids produced by microbial cultures.

←

FIGURE 10.4 Effect of the de Nijs *et al.* [5] treatment on column activity. Column, 15-m SE 52 programmed from 140° to 280°C at 6°C/min. Methyl esters of bacterial acids, standard mixture. Severe adsorption of the more active compounds is evident after extensive use (top). Bottom: behavior following an overnight deactivation [9].

1. Methyl undecanoate
2. Methyl-2-hydroxy decanoate
3. Methyl docecanoate
4. Methyl tridecanoate
5. Methyl-2-hydroxy dodecanote
6. Methyl tetradecanoate
7. Methyl-12-methyl tetradecanoate
8. Methyl pentadecanoate
9. Methyl-2-hydroxy tetradecanoate
10. Methly-3-hydroxy tetradecanoate
11. Methyl-9-hexadecenoate
12. Methyl hexadecanoate
13. Methyl-14-methyl hexadecanoate
14. Methyl-DL-*cis*-9,10-methyl hexadecenoate
15. Methyl hepatdecanoate
16. Methyl-2-hydroxy hexadecanoate
17. Methyl-*cis*-9-octadecenoate
18. Methyl octadecanoate
19. Methyl nonadecanoate
20. Methyl arachidate
21. Methyl heneicosanate

In this particular case esters of saturated and unsaturated acids continued to produce well-formed peaks even after extensive use, but the hydroxy fatty acid esters exhibited malformed peaks that tailed to an unacceptable degree. This gradual increase in column activity may have been due to high-temperature exposure and/or the action of the materials injected. The bottom of Figure 10.4 shows that same column following several hours of the de Nijs *et al.* [5] deactivation treatment [9]; note that the retention behavior has not been affected. Some residual tailing of the more active compounds is still evident; this may be due to active sites in the inlet (see Section 4.4) or in the flame jet.

In other cases column deterioration may take the form of physical displacement of the liquid phase, which can be localized or general. Where physical displacement has been the result of the injection mode or the sample composition, it is usually localized and associated with the injection end of the column. Even a single globule of displaced liquid phase can cause a notable loss in column efficiency [2]. Nonvolatile particulate matter contained in the injected samples or crumbs of silicone rubber from the septum may be blown into the front end of the column. A glass shard from column trimming operations or scrappings from the mounting ferrules may become lodged in the column. Any materials of this type can cause a severe loss in efficiency. Sometimes this is related to liquid phase drawing up around that particle, creating thickened patches of liquid phase that have the same deleterious effect. With transparent columns, localized defects of this type are usually visually apparent, although a low-power magnifying glass may help in the diagnosis. Two remedies are possible: The offending section of column can be cut off and discarded, or it can be cleaned.

If the problem is restricted to two or three coils (i.e., ~1 m) of a column of at least 30-m length, the first solution is usually preferred. That short section can be discarded, and the change in column length will be so slight that it is probable that no change in peak dispersion or elution order (Chapter 8) will be observed.

In cases where residues of dirt have accumulated in the first portion of the column and it seems undesirable to remove that section, Grob [10] suggests that it be cleaned. This is accomplished by washing the first 50–120 cm of the column with an appropriate solvent and then rinsing the washed portion with 0.2% Carbowax 1000 solution.

Where deterioration has occurred throughout the column to such a degree that rejuvenation, either by removal of a column segment or by washiⁿg of a localized area is improbable, some degree of usefulness can occasionally be restored by washing the entire column. It has been well established that a film deposited on a surface cannot be completely removed by washing with solvents in which it is soluble, but that a residual layer remains on the surface, even after cleaning under turbulent conditions [11-13]. A similar phenomenon can be observed in WCOT columns; occasionally columns that have beaded up and lost efficiency can be given a succession of solvent rinsings to remove the globules and excess liquid phase, and the very thin film that sometimes remains may produce a low-capacity, high-efficiency column. A similar finding was recently reported by Ryba [14]. This phenomenon is more frequently encountered with silicone-type liquid phases, but it is not unknown with polar liquid phases.

References

1. Thizon, M., Eon, C., Valentin, P., and Guiochon, G., *Anal. Chem.* **48**, 1861 (1976).
2. Cramers, C. A., Wijnheymer, F. A., and Rijks, J. A., *HRC&CC* **2**, 329 (1979).
3. Grob, K., Grob, G., and Grob, K., Jr., *Chromatographia* **10**, 181 (1977).
4. Jennings, W. G., *Chromatographia* **8**, 690 (1975).
5. de Nijs, R. C. M., Franken, J. J., Dooper, R. P. M., and Rijks, J. A., *J. Chromatogr.* **167**, 231 (1978).
6. Grob, K., *Chromatographia* **10**, 625 (1977).
7. Sandra, P., Verzele, M., and Verstappe, M., *HRC&CC* **2**, 288 (1979).
8. Blumer, M., *Anal. Chem.* **45**, 980 (1973).
9. Mooney, D., and Stern, L., personal communication (1979).
10. Grob, K., Jr., *HRC&CC* **1**, 307 (1978).
11. Anderson, R. M., Satanek, J. J., and Harris, J. C., *J. Am. Oil Chem. Soc.* **36**, 286 (1959).
12. Bourne, M. C., and Jennings, W. G., *Food Technol.* **15**, 495 (1961).
13. Bourne, M. C., and Jennings, W. G., *Nature (London)* **197**, 1003 (1973).
14. Ryba, M., *Chromatographia* **9**, 105 (1976).

CHAPTER 11

COLUMN SELECTION

11.1 General Comments

Selecting a liquid phase for a particular analysis is not a simple task, nor are good guidelines always available. Over the years the practicing chromatographer has accumulated experience and developed an "intuitive feel" for liquid phases that will yield better separations for specific compounds, but the reliability of this acquired sense leaves much to be desired. Two factors affect the separation of components; one is the efficiency—i.e., the number of theoretical plates—delivered by the system, and the other is the relative retentions of those components. Hence the results of other investigations on similar samples are not necessarily a good guide to liquid phase selection. If the cited work used a lower-efficiency system (e.g., packed columns), the liquid phase was necessarily selected for its large α values. With a higher-resolution system much better separation can be achieved even with a less retentive liquid phase, on which the compounds would exhibit shorter analysis times (*vide infra*).

Several points may influence the decision as to which liquid phase should be used in the first trial separation of a particular sample; results of that analysis may then provide additional guidance. One of the first points to consider, if known, is the nature of the sample and the components to be separated. Silylation reaction mixtures should not be used with liquid phases that can react with excess reagent (e.g., Carbowax 20 M). With samples containing low-molecular-weight alcohols, the deactivation treatment of some apolar columns such as SE 30 and SP 2100 may suffer. Although this property can be restored (Section 5.6), tailing would usually be less of a problem with a more polar liquid phase. The separation of lower-boiling sample components will require lower initial column temperatures, but these must of course be above the lower temperature limit of the liquid phase. Carbowax 20 M performs very poorly below 65°C, whereas SP 2100 is usable at 0°C. If the sample contains higher-boiling components, a liquid phase with an elevated temperature limit is desirable unless one can sacrifice efficiency by resorting to extremely high carrier gas velocities or is willing to accept long analysis times and broad, low-intensity peaks. [Alternatively, columns with very thin films of liquid phase can be employed (*vide infra*).] Splitless or on-column injections usually cause a lower degree of column degradation with gum-type liquid phases, such as SE 30, SE 52, or SE 54.

For the first separations of a mixture of unknown composition, the experienced analyst usually prefers to utilize a very short (3–10-m) apolar high-temperature column. Good choices are SE 30 and OV 101 (or SP 2100); SE 52 and SE 54 are also usually suitable. We are normally interested in two types of information from this injection: the degree of component dispersion that is obtained with the apolar column and whether the sample contains high-boiling components. These questions can normally be answered with a single rapidly programmed run. The sample is injected at some suitable temperature (50°–80°C) and the column is programmed to its upper limit (preferably 280°C) and held for a suitable time (30 min). This gives a better picture of the full range of components and makes it more unlikely that higher-boiling constituents that may be well removed from the major portion of the peaks will be missed in subsequent analyses. The presence of high-boiling components may require the use of short or thin-film columns and/or high-temperature liquid phases, preliminary sample fractionation, or advanced techniques such as heart cutting (Chapter 9).

With samples that do not contain high boilers, the resolution achieved on the short apolar column will almost surely be far from satisfactory, unless the sample is a very simple mixture. This should, however, be proven by rerunning the sample on that same short column under isothermal conditions or a lower program rate; in some cases, satisfactory separations will be achieved even with the short apolar column. Where the chromatogram consists of a tight group of poorly resolved low and intermediate boiling solutes, the next separation should utilize a longer apolar column. If the longer apolar column achieves more separation than is required, then the analysis time can be shortened. This can be done by employing higher gas velocities or higher temperatures (or a faster program rate); alternatively, a shorter column can be substituted.

The phase ratio of the column should also receive some attention. Columns with liquid film thicknesses of $d_f = 0.1$-1.0 μ are available; although these values are not always specified by the column supplier, they can be estimated from a knowledge of distribution constants, retention behavior, and Eq. (1.10) and (1.11). Hawkes has published the distribution constants of a number of alkanes in several liquid phases [1-3]. For tetradecane on SE 30 at 150°C, for example, $K_D = 462$. Combining and rearranging Eq. (1.10) and (1.11),

$$d_f = kr_0/K_D \qquad (11.1)$$

If under the preceding conditions C_{14} exhibits a partition ratio $k = 4.0$ in a column of 0.25 mm diameter, $d_f = 0.5$ μ; if $k = 0.8$, $d_f = 0.1$; if $k = 8$, $d_f = 1$ μ. (The effect of these different film thicknesses on resolution is discussed in Section 11.4.)

For columns with d_f values of 0.25 μ and higher, it is usually impractical to resort to column lengths in excess of 50-60 m; peaks that are unresolved on a 50-m column usually remain unresolved on a 100-m column. With d_f values in the neighborhood of 0.1 μ, columns of 100 m may be required to compensate for the low partition ratios, unless low (or subambient) temperatures can be used (see Section 11.4). Columns longer than 100 m are in most cases not practical.

In those cases where the separation on the longer apolar column is still not satisfactory, it is usually unwise to use columns that are still longer than those described above. The logical route at this point is to change to a more polar liquid phase, one in which the relative retentions of solutes will be higher. This is, however, a step that many investigators take with some reluctance, because the polarity of the liquid phase is (in a rough sense) inversely proportional to the stability of the column (Table 11.1). The more polar liquid phases usually exhibit lower temperature limits and higher bleed rates, have lower coating efficiencies, and may require the installation of driers and oxygen scrubbers on the carrier gas lines. Most analysts prefer to use the more stable apolar silicone liquid phases wherever possible. With these thoughts in mind, a Carbowax 20 M column, 25–30 m, would be a logical next choice. Either a shorter column and/or a less polar liquid phase could then be employed if that separation were greater than required; conversely, a longer column (30–60 m) and/or a more polar liquid phase should be used to increase component separation.

Even so, complete separation is not always achieved on a single liquid phase. Yabumoto used high-resolution columns ($n >$ 250,000) on flavor essences of cantaloupe and found that neither Carbowax 20 M nor SE 30 achieved complete separation of all components [4]. On Carbowax 20 M both isobutyl acetate and methyl 2-methylbutanoate exhibit an I of 1008, and 2-methylbutyl acetate and 3-methylbutyl acetate an I of 1119. On an SE 30 column these can be separated, as the first pair have retention indices of 766 and 772, respectively, and the second pair 872 and 870, respectively, on this latter liquid phase. However, pentyl acetate and isobutyl butyrate, which were well resolved on Carbowax 20 M ($I = 1169$ and 1158, respectively), cochromatograph on SE 30 ($I = 905$ and 906, respectively).

The presence of asymmetrically tailing peaks may indicate some special problems. If the defect is due to an acidic function, use of the FFAP or SP 1000 liquid phase will usually give well-formed peaks. If, on the other hand, the tailing is caused by a basic function, incorporation of trace amounts of KOH in the coating solution may prove useful. The use of a suitable test mixture (Section 5.6) can yield some useful information in these cases. Apolar columns will usually exhibit some degree of tailing with alcohols, especially after considerable use or exposure to higher temperatures. Their deactivation can usually be restored by a variety of treatments (Section 5.5); that based on the use of Carbowax 20 M decomposition has proved most effective in the author's hands.

TABLE 11.1

Comparative Properties of Selected Liquid Phases

	Low				High
Upper temperature limit	Polyester		PEG 20 M, SP 1000, FFAP	SP 2250, OV 17 SP 2340, Silar 10	SE 30, SE 52, SE 54, OV 101, SP 2100
Resistance to water[a]	Polyester		PEG 20 M, SP 1000, FFAP	SP 2250, OV 17 SP 2340	SE 30, SE 52, SE 54, OV 101, SP 2100
Resistance to oxygen	Polyester, SP 2340, Silar 10		PEG 20M, SP 1000, FFAP	SP 2250, OV 17	SE 30, SE 52, SE 54, OV 101, SP 2100
Resistance under splitless injection	Polyester	SP 1000, FFAP, SP 2340	SP 2250, OV 17	SP 2100, OV 101	SE 30, SE 52, SE 54
Coating efficiency	Polyester	SP 2340, Silar 10	SP 2250, OV 17	PEG 20M, FFAP	SE 30, SE 52, SE 54
Polarity	SE 30, SE 52, SE 54, SP 2100, OV 101	SP 2250, OV 17	PEG 20m, SP 1000, FFAP	SP 2340, Silar 10	Polyester

[a] Resistance to water (and probably several other properties) may be influenced by the method of column coating, among other things. These ratings are for columns coated by the heated inlet tube modification of the Golay technique. Carbowax 20 M on columns containing a well-prepared barium carbonate deposit seem highly resistant to water, whereas the apolar silicones on this same substrate are much more susceptible to water.

The examples cited in Chapter 15, which are largely limited to separations achieved in open tubular glass columns, can also prove helpful in selecting a suitable column for specific separations. It is clearly evident, however, that although column efficiency is of paramount importance, the polarity of the liquid phase also plays a crucial role.

11.2   The Rohrschneider Concept of Polarity

The desirability of a more rational concept of liquid phase polarity has been marked by many workers, and one of the most logical approaches toward this goal was that of Rohrschneider [5]. He theorized that if a solute were dissolved in a completely nonpolar liquid phase, it would be affected only by nonpolar or dispersive forces, because there were no polar forces (those responsible for induction, orientation, charge transfer, or hydrogen bonding) to interact with it. Squalane (a highly branched saturated hydrocarbon, $C_{30}H_{62}$) was selected as his standard nonpolar liquid phase, and he used Kováts indices (Section 7.3) to describe solute retentions.

A solute RX (of hydrocarbon chain R and functional group X) will exhibit its lowest I in squalane; in any other liquid phase some degree of polarity exists, and these polar forces interact with RX to a greater degree than they do with the n-paraffin hydrocarbons on which I is based. This difference in retention behavior, which is a measure of the increased polarity of the second liquid phase over that ofsqualane, he termed ΔI:

$$\Delta I = I_{\text{polar phase}} - I_{\text{squalane}} \qquad (11.2)$$

Five test compounds were selected to represent a range of functional groups with which the polar forces of a liquid phase could interact: benzene, ethanol, butanone-2, nitromethane, and pyridine. The increase in the retention index of each compound over that exhibited in squalane, divided by 100, yielded the Rohrschneider constants x, y, z, u, and s, respectively. The Rohrschneider constants for Carbowax 20 M, for example, were established as shown in Table 11.2

The Rohrschneider constants x, y, z, u, and s therefore give a qualitative and quantitative indication of the selectivity of a liquid

TABLE 11.2

Compound	$I_{20\,M}$	$I_{squalane}$	$\Delta I/100$	Constant
Benzene	967	649	3.18	x
Ethanol	917	384	5.33	y
Butanone-2	912	531	3.81	z
Nitromethane	1159	457	7.02	u
Pyridine	1199	695	5.04	s

phase. Liquid phases with large x values would be expected to exercise a greater retardation of double-bonded compounds; alcohols would exhibit longer retentions in liquid phases with large y constants.

Rohrschneider reasoned that this selectivity was occasioned not only by properties of the liquid phase that interact with the solute, but also by properties of the solute that interact with the liquid phase. He proposed that

$$\Delta I = ax + by + cz + du + es \qquad (11.3)$$

where a, b, c, d, and e are solute forces that interact with liquid phase forces x, y, z, u, and s, respectively. We have just seen how Rohrschneider constants for a liquid phase are determined. By determining ΔI for a given solute on five different liquid phases whose x, y, z, u, and s constants are known, five simultaneous equations [of the form shown in Eq. (11.3)] can be obtained, which one can solve for the five unknowns a, b, c, d, and e.

McReynolds later proposed slight modifications of this method, but the concept remains the same. McReynolds constants (Appendix II) are widely used today to compare the polarities of different liquid phases, but they give no direct indication as to which liquid phases are best suited to the separation of compounds with similar retention characteristics.

11.3 Other Methods of Selection

Liquid phase mixtures have been shown to possess predictable Rohrschneider (or McReynolds) constants that are intermediate values, and there have been suggestions that liquid phases might be tailor mixed to achieve specific Rohrschneider constants for a given separation [6]. Based on these concepts, Weiner and Parcher [7] described a method of factor analysis to select liquid phases for a limited set of data.

One of the more interesting proposals for column selection was that of West and Hall [8]. They established a simple relationship, based on Kováts retention indices, that took into account column efficiency and liquid phase selectivity to achieve any desired degree of separation of two closely eluting peaks.

Laub and Purnell proposed a novel method for selecting liquid phase mixtures to achieve optimum separation of known [9] and unknown [10] mixtures. The method establishes what volume fractions of two (or more) liquid phases will achieve the highest relative retentions (α values) for all components of a mixture, how many peaks will result, and what the elution order will be. Although its greatest utility is for packed columns because of their inherently lower efficiencies, it could be useful for particularly difficult separations with open tubular columns (Section 9.4).

11.4   Role of Column Efficiency

In selecting a liquid phase for specific separations, attention must also be given to column efficiency. We have seen (Section 1.3) that the resolution of two components is a function of their relative retentions α, the column efficiency n, and their partition ratios k. Suppose, for example, we are interested in separating two components whose α values are 1.056 on the methyl silicone liquid phase SE 30, and 1.194 on the Carbowax 20 M liquid phase. Equation (1.15) indicates that baseline resolution ($R = 1.5$) will require approximately 13,000 effective theoretical plates on an SE 30 column, and about 1400 effective theoretical plates on a Carbowax 20 M column. The individual working with 6-ft $\times$ $\frac{1}{4}$-in. packed columns has no choice; he is restricted to Carbowax 20 M because he requires the large α value to compensate for the low n. But 13,000 effective plates is no barrier to WCOT column; this can frequently be achieved with a column only 4–5 m long. Because of the much higher efficiencies of these columns, many of the restrictions we have observed in the past must be reexamined. Liquid phases of relatively low selectivity will frequently produce excellent separations in a fraction of the time required by the more retentive liquid phases.

Thin-film columns ($d_f \leq 0.1\ \mu$) are sometimes recommended for the separation of higher-boiling or thermally labile substances, because the lower partition ratios of the thin-film columns result in shorter analysis times. Also, column efficiencies (in terms of theoretical plates) are usually much higher for thin-film columns. However, as shown by Eq. (1.14) resolution is directly proportional to the partition ratios; hence (at any given temperature) thin-film columns require more theoretical plates to achieve the same degree of resolution that a thicker-film column attains with lower plate numbers.

It is possible to construct very thin film columns that have very high plate numbers but that are incapable of achieving much in the way of separation, unless column temperatures are lowered so that partition ratios are larger. Instruments that are capable of operation at low temperature can, by adjusting column temperature, utilize film thicknesses of $0.1–0.25\ \mu$ to their full advantage. The vast majority of chromatographs in active use, however, have difficulty in maintaining control of oven temperatures below 50°C. This has created a limited interest in thick-film columns. Johansen [11] demonstrated baseline separation of ethane and methane at 69°C on a column with $d_f = 0.9\ \mu$, in spite of the fact that the number of theoretical plates was relatively low (Figure 5.22). Alternatively, the same compounds can be separated on a conventional column $d_f \simeq 0.3\ \mu$ at subambient temperatures [12]. In most cases a film thickness of $0.25\ \mu$ seems ideal; depending on the coating technique used, columns of high plate numbers can be produced, the C_L term of the van Deemter equation is not yet limiting (Section 8.6), and partition ratios are reasonable. At this film thickness, temperature (or program rate) and/or column length can be varied to achieve the required degree of resolution commensurate with reasonable analysis times and elution temperatures.

References

1. Butler, L., and Hawkes, S., *J. Chromatogr. Sci.* 10, 518 (1972).
2. Kong, J. M., and Hawkes, S., *J. Chromatogr. Sci.* 14, 279 (1976).
3. Millen, W., and Hawkes, S., *J. Chromatogr. Sci.* 15, 148 (1977).

4. **Yabumoto, K.,** Ph.D. Thesis, Univ. of California, Davis, 1976.
5. **Rohrschneider, L.** *J. Chromatogr.* **22,** 6 (1966).
6. **Jennings, W. G.,** *Chem., Mikrobiol., Technol. Lebensm.* **1,** 9 (1971).
7. **Weiner, P H.,** and **Parcher, J. F.,** *J. Chromatogr. Sci.* **10,** 612 (1972).
8. **West, S. D.,** and **Hall, R. C.,** *J. Chromatogr. Sci.* **14,** 339 (1976).
9. **Laub, R. J.,** and **Purnell, J. H.,** *Anal. Chem.* **48,** 799 (1976).
10. **Laub, R. J.,** and **Purnell, J. H.,** *Anal. Chem.* **48,** 1720 (1976).
11. **Johansen, N.,** *Chromatogr. Newsl.* (to be published).
12. **Rooney, T. A.,** personal communication (1979).

CHAPTER 12

SAMPLE PREPARATION

12.1 General Considerations

While gas chromatography has permitted tremendous advances in the area of volatile analysis, it has been subjected to a high degree of abuse, and the results of few other processes have been so frequently and so flagrantly overinterpreted. Far too often the chromatogram is assumed to be an accurate representation of the composition of the starting material. This assumption fails to take into account the fact that with some notable exceptions, most substances require extensive preliminary treatment to prepare an extract or essence suitable for gas chromatographic analysis. The methods used for this sample preparation can have a profound effect on the composition of the sample injected, and not all substances that are finally injected are stable to the gas chromatographic process: Some components of the original sample never get into the inlet, some never get out of the inlet, and some fail to survive passage through the column. The use of unpacked glass columns in all-glass system rectifies some but not all of these problems.

Some essential oils and samples such as relatively concentrated chemical reaction mixtures or mixtures of low-boiling petroleum fractions may be amenable to injection into the chromatograph per se, but even here a preliminary isolation or fractionation treatment can raise the concentration of selected trace constituents to a detectable level. The investigator who is interested in studying the composition of a dilute vapor system (e.g., air, stack gas, breath analysis) or samples containing large amounts of water, alcohol, or nonvolatile materials (including most biological products) is faced with a new set of problems. Injections containing large amounts of water, alcohol, or other strongly adsorbed materials can lead to rapid column deterioration, and the injection of nonvolatile materials should be scrupulously avoided. Not only can such contaminants lead to column deterioration, but they may also slowly degrade to produce volatiles that cause severe noise problems, or they may react with materials injected subsequently, producing extraneous results. This can be especially troublesome with samples of biological origin. Some method of separating these nonvolatile components must be utilized if the volatile components are to be studied without interference.

We are really faced with two separate problems. One might be termed isolation, and consists of separating the volatiles of interest from interfering substances such as nonvolatile materials, water, or large amounts of ethanol. The other problem is concentration and is concerned with raising the level of materials to be detected to a point at which the restricted sample that can be placed on the column contains detectable amounts of the compounds of interest.

Because of the interrelated facets of these problems, it is quite impossible to select any one isolation or concentration procedure as being uniformly satisfactory. It is instead necessary for the investigator to decide, with attention to his system and what he intends to study, which of the methods available is most satisfactory for his sample and objectives or which could be adapted to achieve his goals.

12.2   Headspace Concentrates versus Total Volatile Analysis

Classical methods for the isolation and concentration of volatile components have been well reviewed by Weurman [1]. Although he gave his primary attention to food odors, many ideas he raised are applicable to other systems, both biological and nonbiological.

He pointed out that the total volatile content of a substance is related to the composition of the headspace volatiles from that substance by the distribution of those volatiles within and the physical state of the substance. Figure 12.1 illustrates this point. The diagrams show, in the upper part, two biological products, basically aqueous, that contain polar and nonpolar volatiles, nonvolatile insoluble lipids, and nonvolatile solutes that might include sugars, salts, amino acids, and proteins. In addition, the product depicted in the left half of the figure contains undissolved solid particulate matter, and some of the solutes, both volatile and nonvolatile, are shown absorbed to the surface of this material. Where sufficient lipid material is available, as in the product in the right-hand portion of the figure, all nonpolar volatile materials is considered to be dissolved in these globules of lipid. In the left-hand example, however, a limited amount of lipid is available, and some of the nonpolar volatile material is left suspended in the aqueous phase. The circles in the lower portion of the diagram depict the distribution of the various solutes in a headspace sample and following distillation. Even though the total amount of volatiles is qualitatively and quantitatively the same in both instances, the relative vapor pressures of the components, and hence their concentrations in the "head space," are different because the distribution of the components is different.

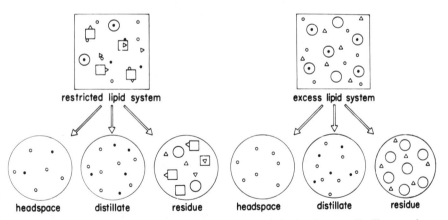

FIGURE 12.1 Effect of sample composition of the headspace, distillate, and residue. ○, Volatile polar solutes; •, volatile nonpolar solutes; ▲, nonvolatile solutes; ◯, undissolved lipid materials, □, solids.

12.3   Direct Headspace Analysis

The direct injection of a headspace gas has an appealing simplicity, and for some samples can yield satisfactory results [2, 3]. Unfortunately, only those substances whose vapor pressures are sufficiently high that they are present in the headspace in amounts large enough to activate the detector will produce peaks. Although a larger headspace injection would contain greater amounts of less volatile materials, large injections are not consistent with narrow solute bands and sharp peaks, but produce instead broad overlapping peaks and poor resolution. Some investigators have attached a cold trap as the inlet to the column, or chilled a short section of the front of the column (i.e., cold trapping, Section 4.2). The noncondensable gas passes through, while the condensable volatiles concentrate in the trap or on the chilled column front. When this trapping area is then heated, the chromatographic process is begun. Two problems are immediately apparent: It is difficult to design a precolumn trap so that its contents are rapidly transferred to the column at the low flow rates used in these high-resolution systems, and for samples containing water vapor, the major volatile recovered is usually water. With our present methods of detection, relatively few systems can be adequately explored by simple headspace techniques, but in those cases where one is interested in lower-boiling volatiles, the method can prove extremely useful (*vide infra*) (Figures 15.16 and 15.36). Certain precautions may prove necessary; glass syringes used to inject the sample sometimes exhibit a demand capacity and selectively abstract sample components, and some disposable plastic syringes contribute peaks of their own. The exploration of a variety of headspace sampling techniques and their application to a range of food products was the subject of a recent symposium [4].

Parliament [5] described a method of utilizing cocondensation to trap effluent fractions from packed-column GC in refluxing carbon tetrachloride. Jennings applied an adaptation of this to trap headspace volatiles in refluxing Freon [6] (Figure 12.2). An automated technique for introducing headspace samples from water, flavor materials, and crude oils onto glass capillary columns has been described by Kolb *et al.* [7]. The apparatus utilized by the latter authors is shown in Figure 12.3. Rapp conducted a sweeping gas through alcoholic beverages to a Freon system to

FIGURE 12.2   A Freon cocondensation unit, patterned after a cocondensation technique described by Parliament [5]. When sample-containing vapors are slowly passed through the column of refluxing Freon, entrained volatiles are forced to the flask. The Freon is then allowed to evaporate. (After Jennings [6].)

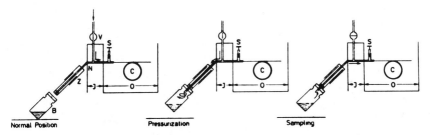

FIGURE 12.3   Headspace sampling apparatus described by Kolb *et al.* [7]. Carrier gas is introduced into the sample-containing vial until its pressure equals that of the column head pressure. A valve then interrupts the carrier gas flow and gas from the pressurized vial is transferred to column. After 5 sec the valve restores carrier gas flow, stopping the injection. Typical chromatograms are shown in Figures 15.17 and 15.37.

recover samples of selectively enriched headspace volatiles (low in water and ethanol) [8].

12.4 General Considerations on Methods of Isolation and Concentration

Most methods for accomplishing the isolation or concentration of volatiles involve crystallization (including that of water), distillation, extraction, or adsorption steps. Any of these steps can cause quantitative changes in the ratios of the recovered materials, and some can cause qualitative changes that may take the form of the generation of new volatiles or the disappearance of existing ones.

Many materials require some type of maceration or cutting as a first step in the isolation of the volatile fraction. Especially with biological substances, great care must sometimes be exercised to avoid chemical or enzymatic changes during these steps. A critical point, which was also emphasized by Weurman [1], is that errors committed in the early stages of isolation and concentration can never be corrected at any later stages of analysis.

12.5 Distillation

Distillation is the step most usually employed to isolate the volatile from nonvolatile materials; it is sometimes preceded and sometimes followed by a concentration step involving adsorption, extraction, or freezing (*vide infra*). Distillation may take the form of a simple single plate still operated at atmospheric or reduced pressure or a multiplate unit utilizing reflux stripping; in either case the material undergoing stripping is subjected to a relatively rigorous treatment for a relatively long period of time. Reduced pressure climbing film, falling film, and rotary film evaporators that have become available in more recent years offer several advantages and are less apt to produce artifacts because the sample contact time with the heated surface is much shorter. A number of other distillation procedures, including closed and recycling systems, were discussed by Weurman, and later by Teranishi *et al.* [9]. One worthy of special mention is a simultaneous distillation-extraction unit described by Nickerson and Likens [10]. This device, shown in Figures 12.4 and 12.5, is also capable of operation under reduced pressure. Excluding the method just described, most of these procedures result in a dilute aqueous distillate from which the volatiles are subsequently extracted or adsorbed; a preliminary concentration of this distillate is sometimes advantageous (*vide infra*).

The addition of water to an anhydrous sample (e.g., fats or oils) or to one containing a limited amount of water may sometimes be undesirable. In such cases carbon dioxide [11], nitrogen [12], or air [13, 14] may be used to sweep the volatiles to a cold trap, or adsorbant, or an absorbant (*vide infra*). High-vacuum transfer can also be very useful for isolating volatiles from a variety of materials without resorting to gas or steam distillation. Figure 12.6 shows a simple device for accomplishing this.

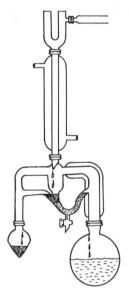

FIGURE 12.4 Schematic of the Nikerson-Likens simultaneous steam distillation-extraction apparatus shown in Figure 12.5. It is advantageous to mount the assembly on a separate rack, with a central pivot point, so that the side arm levels can be precisely adjusted by tilting the entire apparatus.

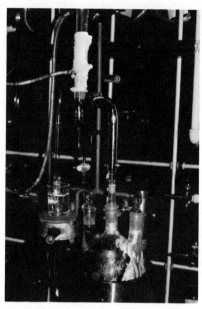

FIGURE 12.5   A modified distillation–extraction apparatus. The distillate from the large sample flash (right) cocondenses with the constantly recycled solvent (left). The less dense solvent is returned through the left-hand side arm to the solvent reservoir, whereas the aqueous residue returns to the sample flask through the right-hand side arm. (See Nikerson and Likens [10].)

12.6   Freeze Concentration

The extraction of large volumes of aqueous material will require large volumes of extraction solvents (or elaborate recycling systems, which pose their own problems, *vide infra*). Large volumes of solvent are undesirable from many standpoints, but of especial concern is the necessity for subsequently removing the major portion of the solvent (and a portion of the extracted volatiles) prior to analysis, and the resultant concentration of solvent-borne impurities that can in some cases then dominate the analysis. It is frequently desirable to reduce the volume of the aqueous sample prior to extraction or adsorption. Under carefully controlled conditions it is possible to freeze out essentially pure water, so that the concentration of other materials is proportionately higher in the reduced volume of unfrozen liquid. The method is attractive, in that the sample is subjected to very mild conditions, and

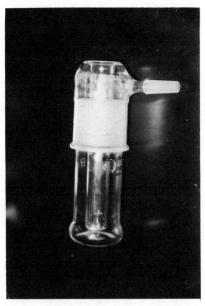

FIGURE 12.6 A simple cold-finger trap for operation at high vacuum. The sample—which can range from a food product to crude oil or tar—is placed in the large chamber, and the central cold finger is filled with a coolant—dry ice or liquid nitrogen. Vacuum is applied at the side arm, and the sample may be heated by placing the apparatus on a hot plate.

the apparatus requirements are simple. Batches of solution may be placed in a beaker subjected to continuous stirring at subfreezing temperatures, or in a round-bottomed flask that is continuously rotated while half submerged in a freezing bath. To avoid occlusion and entrapment, freezing should occur slowly and under equilibrium conditions. Hence both the rate of stirring (or rotation) and the temperature of the bath are critical. Concentrations of five- to fortyfold, equivalent to 80–95% water removal, have been reported [15, 16].

12.7 Extraction

Extraction is usually—but not always—preceded by a separation step such as stream distillation. The extracting liquid is normally a nonpolar organic solvent, and a wide range of these have been used. If a low-boiling solvent is used for the extraction, the major portion must be removed by fractional distillation or evaporation prior to analysis. If a high-boiling solvent is used, the solvent is subjected to stripping by distillation, usually at reduced pressure, to recover the more volatile compounds. This latter approach is usually more profitable when the investigator is interested in the isolation of lower-boiling volatiles. Salts such as sodium chloride or sodium sulfate are frequently used to increase competition for the aqueous phase and shift the distribution coefficients of the desired compounds to favor the organic solvent. The effect of salt addends, however, is sometimes more complicated than is usually realized, and their use should be considered rather carefully. In some cases salt forces solutes out of the aqueous phase, but they are then more prone to appear in the headspace than in the organic solvent used for extraction.

The extraction solvent is usually selected with a view to its extraction efficiency, its inertness, and its boiling point. In the extraction of aqueous steam distillates, some investigators prefer ethyl ether because of its extraction efficiency, although pentane or isopentane may offer an advantage for fermentation products; the hydrocarbons have a lower overall extraction efficiency, but the lower-molecule-weight alcohols that would dominate an ether extract are largely left unextracted in the water phase. Ethyl ether is prone to form peroxides, and its use as an extraction solvent by the unwary can lead to artifact formation. In general, the lower the boiling point of the extraction solvent, the less serious are the losses of lower-boiling volatiles during the final concentration step. Even so, attention must also be directed to the volume of solvent removed; even volatiles with appreciably higher boiling points contribute to the vapor pressure of the system, and a loss in volatiles, proportional to their concentrations and their vapor pressures, is to be expected during concentration. Final concentration is sometimes achieved by directing a stream of nitrogen into a test tube containing the concentrated extract. As the solvent evaporates, the extract becomes cold, and impurities in the gas will concentrate and complicate the analysis. Under extreme conditions moisture in the atmosphere may condense in the tube. Gas-borne impurities can usually be removed by first passing the gas through molecular sieves (*vide infra*).

Fluorocarbons, commercially available as the freons, have been used as extraction solvents, as has ethyl chloride. Liquid carbon dioxide in a pressurized system [17] has also been utilized. Jennings [6] described a pressurized system permitting the use of liquid carbon dioxide with a standard glass Soxhlet extractor (Figure 12.7). The unit, which can utilize dry ice as a carbon dioxide source, has the advantage of achieving high extraction efficiencies under an inert atmosphere and at low temperature. It has been employed for simple extractions as well as the recovery of volatiles from adsorbants such as activated carbon, XAD resins, or porous polymers. The carbon dioxide is allowed to dissipate at subzero

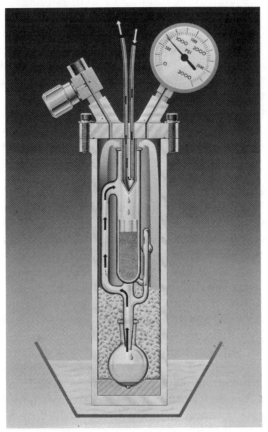

FIGURE 12.7 High-pressure Soxhlet extractor, usable with liquid carbon dioxide to yield solvent-free extracts. See text for details. (From Jennings [6]. Photograph courtesy of J&W Scientific, Inc.)

temperatures, yielding a solvent-free extract for analysis. In all such cases, where one is dealing with a system of unknown composition, it becomes critically important to include a blank so that it is possible to ascertain which of the peaks in the final analysis were contributed by components of the system.

12.8   Zone Refining

Zone refining (or zone melting) offers some real advantages in the area of sample preparation, in that extreme concentration should be possible under very mild conditions. Benzene, with a freezing point of 5.5°C, is a solvent well suited to this technique. Huckle [18] achieved a 3000-fold concentration of a raspberry juice with this method, which could also be applied to steam distillates. It depends on the principle that a solvent freezing slowly under equilibrium conditions tends to exclude impurities and freeze as pure solvent. Huckle filled a glass tube with benzene, which was then frozen. Utilizing an apparatus simlar to that shown in Figure 12.8, several evenly spaced "zones of melt" were passed along the tube by narrowly confined and slowly moving peripheral heaters. The energy of the heaters and the temperature of the air surrounding the nonheated zones of the tube must be carefully adjusted so that each melted zone refreezes under equilibrium conditions as the moving heaters pass on. The impurities in the benzene tend to remain with the melt and are eventually swept to the bottom of the tube. After the zones of melt have swept the tube some numbers of times, which may require several days, the entire tube is allowed to freeze and the bottom portion discarded. The carefully purified solvent can then be used as an extracting solvent, recovered, and resubjected to zone refining. The impurities recovered this time are the volatiles of interest, which may be injected at high concentration.

Attractive though the method is, it has not received wide usage. The major problem, at least in our hands, has been related to the fluctuations that occur in the air bath used to attain the freezing temperature. The energy supplied to the heaters can be controlled to within very close tolerances, but as the temperature of the air bath shifts over a 1° or 2° range, the entire tube freezes and no sweeping occurs, or it melts and the entire contents remix.

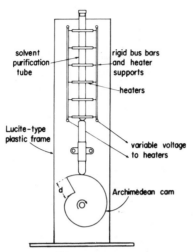

FIGURE 12.8 Typical zone melting apparatus. Heaters are aligned and locked on the rigid bus bars and spaced at intervals equivalent to d on the Archimedean cam. At the end of each cam revolution, the centerline of each zone of melt becomes the centerline of the zone of melt for the next lower heater, giving six continuously moving zones of melt. From 2 to 6 hr may be required for one cam revolution.

12.9 Adsorptive Methods

Activated carbon in the form of granular charcoal has been widely used for the isolation of volatile constituents from a vapor or water stream [17, 19–21]. Although there is a strong possibility of artifact formation in any adsorptive process, no significant degree of rearrangement or degradation has been found by the several workers who investigated this possibility [22–24]. In our own laboratory, essences of Bartlett pear exhibited a slight degree of rearrangement as evidenced by the fact that a small amount of the *trans*:2-*cis*:4-decadienoate esters apparently rearranged to the trans–trans configuration after adsorption on charcoal and storage for eight years at room temperature.

Coconut shell and bone charcoal have both been used, either in pelletized form or as a relatively coarse (e.g., 20–40 mesh) grind. Following the adsorptive step, the charcoal can be subjected to elution without preliminary drying with a variety of solvents, including ethyl ether, pentane, carbon disulfide, or liquid carbon dioxide. Alternatively, it can be lyophillized to remove water, and subjected either to solvent elution or to slowly ascending temperature while under high vacuum. The latter procedure can produce an anhydrous, solvent-free essence. Some degree of selectivity can sometimes be demonstrated, either in the adsorptive or desorptive step [1]. An interesting modification of the charcoal-trapping technique described by Grob [25] was developed by Klimes and Lamparsky [26] to prepare samples of vanilla bean volatiles.

Adsorption chromatography is widely used to effect prefractionation of some complex mixtures. Essential oils serve as a good example. Many of these contain a variety of terpenoids, and their chromatograms are often characterized by three groups of closely associated compounds: the monterpene hydrocarbons, the sesquiterpene hydrocarbons, and the oxygenated terpenes. Because the volatilities and polarities of the compounds comprising these groups differ, multiramp programming can be beneficial in the separation of these mixtures. Alternatively, prefractionations are often used. HPLC offers one approach, but an even simpler technique is widely used. A 15-cm length of 1–2-mm thin-wall Teflon tubing is packed with a suitable adsorbent, such as silica gel, to construct a microchromatographic column. One places 1–2 μl of the essential oil on one end of this, and the column is placed, sample end down, in a stoppered test tube containing ~1 ml of an appropriate solvent; dichloromethane is widely used. As the solvent reaches the top of the column, it is removed and sectioned with a razor blade. Ether can be used to recover volatiles from the individual sections. The monoterprene hydrocarbons are usually close to the solvent front, the sesquiterpene hydrocarbons intermediate, and the oxygenated compounds last.

12.10 Porous Polymer Absorption

Some years ago new materials in the form of porous beads of plastic were developed for use in gel permeation chromatography. Typically they were copolymers of styrene and divinylbenzene or related compounds and were marketed under various trade names (see Appendix III). Because they were analogous to a precoated solid support in that this spherical granular material possessed exposed functional groups, several workers explored their use as gas chromatographic substrates. Results indicated that their utility as column packing materials was limited to a few special cases but that they possessed certain unique properties: water and low-molecular-weight alcohols exhibited very short retentions (Appendix III) and were eluted as symmetrical peaks. This suggested the possibility that the porous polymers might be utilized for sample preparation; large volumes of air or "headspace" gas could be swept through tubes (or traps) packed with a porous polymer. Materials with short retentions, including noncondensable gases, water, and low-molecular-weight alcohols, would pass through the trap in a relatively short time. Other materials would accumulate, and the trapped volatiles could be recovered by backflushing to a cold trap [27, 28]. The method proved quite successful, and has been widely applied to a large number of materials ([2, 13, 14, 29–33]; see Chapter 15 also).

Simple though the process seems, many workers have experienced varying degrees of difficulty. As supplied, most porous polymers are heavily contaminated, and a rigorous conditioning is required. It should also be emphasized that unless the sweeping gas—usually air or high-purity nitrogen—is prepurified, the major volatiles recovered are generally the trace impurities present in that gas. Several methods have been used, but passage through a mixed bed of molecular sieves 4A (or 5A) and 13X is usually adequate. Several porous polymers undergo some degree of degradation when heated in the presence of air or oxygen; conditioning and recovery are best accomplished with an inert gas such as nitrogen.

In most cases traps consist of a small quantity (0.5–20 g) of porous polymer in short lengths of glass tubing (4–15-mm o.d.) between glass wool plugs. Conditioning is usually accomplished by passing prepurified nitrogen through the trap at an elevated temperature. As mentioned previously, it is generally wise to displace the air before raising the temperature; maximum temperatures should be well below the upper limits specified by the supplier. Some of the porous polymers require extensive conditioning and, at least with some batches, the contribution of artifact peaks cannot be avoided; hence blank runs are essential for comparative purposes. In general, those with higher sample-holding capacities (the Porapaks) require more cleanup, contribute more artifact peaks, and are less heat stable than those with more restricted sample-holding capacities (Tenax GC).

Trapping is accomplished with the gas (air or nitrogen) flowing in the same direction used for trap conditioning. Where large quantities of water vapor may be present, condensation is sometimes discouraged by maintaining the trap at a temperature 5°–10°C higher than ambient. The sample is then removed and a development step [34] is sometimes used to flush the last traces of the faster-moving constituents (water, ethanol) out of the trap. Boyco et al. explored the retentions of a range of volatiles on several porous polymers in selecting precolumn parameters useful in the removal of water or low molecular weight alcohols [35]. The trap is reversed, and the sample recovered by backflushing with nitrogen at an elevated temperature. Various "purge and trap" devices are available; in most cases these are merely convenient methods for the application of these techniques. The sample stream is passed through a trapping substrate, thermally desorbed, and cold trapped on column. Solvent elution of the porous polymer has also been suggested [36]; this method appears to work best with the Tenax GC porous polymer (Appendix III). One end of the tube, which is later packed with the trapping agent, is first drawn to a long, tapering capillary. After trapping, a minimum amount of freshly redistilled ethyl ether is percolated through the porous polymer until a small quantity flows to the capillary tip. This is withdrawn in a microsyringe for analysis. The liquid carbon dioxide extractor described earlier (Section 12.7, Figure 12.7) works well with Tenax GC. The adsorbant should be thoroughly cleaned by extensive CO_2 extraction prior to the trapping step.

Samples of polyurethane foam have also been used to entrain microparticulate matter (and volatiles?) from airstreams. The polyurethane trapping plugs are then subjected to solvent extraction to prepare samples for analysis [37]. The method seems particularly applicable to the analysis of airborne pesticides.

References

1. **Weurman, C.,** J. Agric. Food Chem. **17,** 370 (1969).
2. **Jennings, W. G.,** and **Filsoof, M.,** J. Agric. Food Chem. **25,** 440 (1977).

3. Yabumoto, K., Yamaguchi, M., and Jennings, W. G., *J. Food Chem.* **3**, 7 (1978).

4. Charalambous, G., ed., "Analysis of Food and Beverages, Headspace Techniques." Academic Press, New York, 1978.

5. Parliament, T. H., *Anal. Chem.* **45**, 1792 (1973).

6. Jennings, W., *HRC&CC* **2**, 221 (1979).

7. Kolb, B., Pospisil, P., Borath, T., and Auer, M., *HRC&CC* **2**, 283 (1979).

8. Rapp, A., *in* "Applications of Glass Capillary Chromatography" (W. Jennings, ed.) Dekker, New York, 1980. In press.

9. Teranishi, R., Hornstein, I., Issenberg, P., and Wick, E. L., "Flavor Research." Dekker, New York, 1971.

10. Nickerson, G. B., and Likens, S. T., *J. Chromatogr.* **21**, 1 (1966).

11. Honkanen, H., and Karvonen, P., *Acta Chem. Scand.* **20**, 2626 (1966).

12. Nawar, W. W., and Fagerson, I. S., *Food Technol. (Chicago)* **16**, 107 (1962).

13. Tressl, R., and Jennings, W. G., *J. Agric. Food Chem.* **20**, 189 (1972).

14. Jennings, W. G., and Tressl, R., *Chem., Mikrobiol., Technol. Lebensm.* **3**, 52 (1974).

15. Kepner, R. E., van Straten, S., and Weurman, C., *J. Agric. Food Chem.* **17**, 1123 (1969).

16. Shapiro, J., *Anal. Chem.* **39**, 280 (1967).

17. Schultz, T. A., Flath, R. A., Black, D. R., Guadagni, D. G., Schultz, W. G., and Teranishi, R., *J. Food Sci.* **32**, 279 (1967).

18. Huckle, M. T., *Chem. Ind. (London)* p. 1490 (1966).

19. Ralls, J. W., McFadden, W. H., Siefert, R. M., Black, D. R., and Kilpatrick, P. W., *J. Food Sci.* **30**, 228 (1965).

20. Tang, C. S., and Jennings, W. G., *J. Agric. Food Chem.* **15**, 24 (1967).

21. Jennings, W. G., and Nursten, H., *Anal. Chem.* **39**, 521 (1967).

22. Carson, J. F., and Wong, F. F., *J. Agric. Food Chem.* **9**, 140 (1961).

23. Paillard, N., *Fruits* **20**, 189 (1965).

24. Heinz, D. E., Sevenants, M. R., and Jennings, W. G., *J. Food Sci.* **31**, 63 (1966).

25. Grob, K. *J. Chromatogr.* **84**, 255 (1973).

26. Klimes, I., and Lamparsky, D., *in* "Analysis of Foods and Beverages. Headspace Techniques" (G. Charalambous, ed.), p. 95. Academic Press, New York, 1978.

27. Dravnieks, A., and O'Donnell, A., Am. Chem. Soc., 160th Natl. Meet., Chicago, 1970 Pap. No. 2.

28. Jennings, W. G., Am. Chem. Soc., 160th Natl. Meet., Chicago, 1970 Pap. No. 99.

29. Jennings, W. G., Wohleb, R. H., and Lewis, M. J., *J. Food Sci.* **37**, 69 (1972).

30. Uchman, W., and Jennings, W. G., *J. Food Chem.* **2**, 135 (1977).

31. Jennings, W. G., *J. Food Chem.* **2**, 185 (1977).

32. Zlatkis, A., Bertsch, W., Lichtenstein, H. A., Tishbee, A., Shunbo, F., Liebich, H. M., Coscia, A. M., and Fleischer, N., *Anal. Chem.* **45**, 763 (1973).

33. Novotny, M., McConnell, M. L., and Lee, M. L., *J. Agric. Food Chem.* **22**, 765 (1974).

34. Jennings, W. G., Wohleb, R. H., and Lewis, M. J., *Master Brewers Assoc. Am. Tech. Q.* **11**, 104 (1974).

35. **Boyco, A. L., Morgan, M. E.**, and **Libbey, L. M.**, *in* "Analysis of Food and Beverages. Headspace Techniques" (G. Charalambous, ed.), p. 57. Academic Press, New York, 1978.
36. **Shaefer, J.**, personal communication. TNO, Zeist, Netherlands.
37. **Turner, B. C.**, and **Glotfelty, D. E.**, *Anal. Chem.* **49**, 7 (1977).

CHAPTER 13

ANALYSIS OF MATERIALS OF RESTRICTED VOLATILITY

13.1 General Considerations

Guichon [1] pointed out that a major thrust in GC has been pointed toward extending the analytical capability to more complex and less volatile compounds. The routes explored have included the use of columns with larger phase ratios (i.e., thinner films, hence smaller partition ratios), using less retentive liquid phases, and the preparation of volatile derivatives. Schomburg *et al.* [2] emphasized that decomposition of these larger compounds may occur at the temperatures required for reasonable elution times. They suggest the use of hydrogen carrier gas (Section 8.6), careful column deactivation, and the use of low-polarity liquid phases. These and other approaches are discussed in this chapter.

13.2 Chromatographic Considerations

Because of their lower volatilities, analyses of samples containing these larger and/or more complex substances require higher temperatures, both within the column and in the inlet and detector. A significant amount of decomposition can occur in the inlet, and it may be advantageous to reflect that splitless injection techniques can be used at lower inlet temperatures than split techniques, largely because a longer period of time is allowed for the sample to be volatilized and transported to the column (Section 4.3). On-column injection (Section 4.5) exposes the sample to the lowest possible temperatures.

As discussed in Section 8.7, the elution temperature of a substance under any given set of conditions is a function of the program rate (or temperature) and the carrier gas velocity. There are finite limits to column temperature; deactivation treatments suffer, liquid phase bleeding is accentuated, column deterioration is accelerated, and labile sample components decompose. Hence higher carrier gas velocities become even more important in the analysis of high-boiling substances. Because it yields equivalent efficiencies at much higher linear velocities, hydrogen is the preferred carrier. This point has been emphasized by Schomburg et al. [2] and by Grob and Grob [3]. For the purpose under consideration, the carrier gas should be supplied at the OPGV, and where surplus separation efficiency can be sacrificed in the interest of shorter analysis times and lower elution temperatures, significantly higher velocities may be in order.

Apolar liquid phases normally have the highest operating temperatures and provide the shortest retentions. Columns with very low loading (d_f 0.1–0.2 μ) may be beneficial, as partition ratios (and retention times) will be smaller. Very thin film columns, however, achieve lower resolution at equivalent plate numbers (Section 11.4) and may expose active sites, leading to the adsorption and/or decomposition of sensitive compounds. The behavior of some compounds on apolar columns can be critically affected by the deactivation of that column. Most deactivation treatments are adversely affected at elevated temperatures, which can also cause problems: Fused silica columns (Section 2.2) may offer help in this direction, but at this writing only limited data are available on their performance.

Liquid phases or columns with higher temperature limits would appear to hold some promise for higher-temperature analysis, but that promise has yet to be fulfilled. Glass capillaries prepared with what have been described as bonded liquid phases have been reported to possess greater thermal stability [4, 5]. Others imply that their behavior is more due to the nature of the glass used and is comparable to that of siloxane-type phases on borosilicate glass as contrasted to the more alkaline lime-soda substrate offered by soft glass (e.g., [2]). High-temperature liquid phases such as Dexsil 300 have long held appeal for those interested in the analysis of low-volatility compounds, but their behavior in glass capillary columns is usually disappointing. Although Dexsil has a recommended maximum operating temperature of 500°C, it rarely endures in the glass capillary beyond 300°C, and its bleed rate is so high that its utility for GC/MS is limited to ~165° [6]. Poly S-179 [7], a moderately polar liquid phase whose temperature stability extends to 400°C [8], also attracted some interest. The material coats well on glass, but Schomburg found that its high viscosity at temperatures below 200°C caused serious limitations [2]. Our own experiences with this liquid phase indicate that, although it is capable of withstanding higher temperatures, its higher low-temperature limit coupled with its longer retentions negates the advantage of that higher high-temperature limit. With a Poly S-179 column it may be necessary to use a temperature of 350°C to produce a chromatogram equivalent to one obtained on SE 30 or SE 54 at 250°C. Short thin-film columns coated with methyl silicones can be used for extended periods at 280°C and for shorter periods at 300°C, and some users report exposing them to temperature programs as high as 350°C. Gum phases (OV 1, SE 30, SE 52, SE 54) are more stable at high temperatures than are the silicone fluids. Superior results will be obtained with columns prepared from borosilicate (Duran) glass as contrasted with soda-lime (AR) glass, as the more alkaline surface of the latter causes decomposition of the silicones at higher temperatures, manifested by higher bleed rates.

13.3 Derivatization

Molecular complexity, or the presence of certain functional groups, is frequently responsible for lower volatility. As an example, compounds that contain active hydrogen atoms (e.g., amino, hydroxyl, or carboxyl groups) may interact with other similar molecules or with the liquid phase. Compounds of this type are sometimes derivatized to produce a more volatile substance that can be analyzed by gas chromatographic techniques. Fatty acids, for example, are frequently converted to the corresponding methyl esters. Not only do the resultant esters possess a greater volatility, but by eliminating the active hydrogen of the carboxyl group, the possibility of interactions with the chromatographic system are greatly reduced. A number of derivatization reactions as applied to the analysis of food additives was recently reviewed [9], and the subject of chemical derivatization techniques for GC analysis of pesticides was thoroughly explored by Cochrane [10]. Kossa and MacGee discussed pyrolytic methylation, which, because of the temperatures and equipment utilized, seems more nearly akin to derivatization than to pyrolysis. The process involves the decomposition of the N-methylammonium salt of an acidic compound within the injection chamber of the GC (220°–375°C) to form a volatile methyl derivative [11]. General methods for the derivatization of some specific types of compounds are mentioned in Chapter 15.

A prime example of the use of volatile derivatives to extend the advantages of gas chromatographic analysis to nonvolatile materials is to be found in the field of metal analysis, which has been well reviewed [12]. A number of derivatives have been used, but the metal chelates seem best suited to these analytical requirements. Recent efforts [13] have utilized the enhanced sensitivity of selective detectors in the analysis of several metals by glass capillary GC (Figure 13.1).

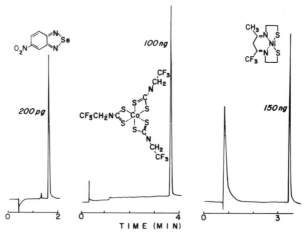

FIGURE 13.1 Chromatograms of selected volatile derivatives of selenium, cobalt, and nickel. N/P detector with a 15- × 0.25-mm column coated with SP 2100.

13.4 Silylation

The development of reliable silylation techniques, which result in the substitution of a trimethylsilyl group for the active hydrogen atom (or occasionally for the metal ion of a salt), has permitted the application of gas chromatography to a much wider spectrum of compounds. Poole and Zlatkis have reviewed trialkylsilylether derivatives other than TMS derivatives in terms of the reactions involved and their applicability to a wide area of analyses [14].

As with the methyl ester example, the substitution of these groups for active hydrogen reduces the polarity of the compound and lessens the tendency to form hydrogen bonds. Hence with compounds that exhibit a strong tendency for hydrogen bonding, the silyl derivative is usually more volatile. The reactivity of the compound has also been reduced, making it less prone to react with other substances or the components of the system.

$$R-OH \xrightarrow[\text{TMCS}]{(CH_3)_2SiCl} R-O-Si(CH_3)_3 + HCl$$

$$R-COOH \xrightarrow[\text{TMSDEA}]{(CH_3)_3Si-N(C_2H_5)_2} R-COOSi(CH_3)_3 + (C_2H_5)_2NH$$

$$2CH_3C \overset{OSi(CH_3)_3}{\underset{NSi(CH_3)_3}{}}$$

$$R-\underset{\underset{NH_2}{|}}{CH}-COOH \xrightarrow[\text{BSA}]{} CH_3-\underset{\underset{\underset{Si(CH_3)_3}{|}}{NH}}{CH}-\overset{\overset{O}{\|}}{C}-OSi(CH_3)_3$$

$$+ 2CH_3C \overset{O}{\underset{NHSi(CH_3)_3}{\diagup}}$$

The reagents for these and other similar reactions are readily available (Appendix IV), and the techniques are simple and straightforward. The reactions are reversible, and in the presence of moisture the parent compound is regenerated from the derivative.

Although this is occasionally advantageous in regenerating a compound after separation, it can also hinder the separation. Some silylated derivatives—amino acids are a good example—can be chromatographed in packed columns. When attempts are made to repeat the analysis in a higher-resolution glass capillary system, not all components survive the separation. Three factors appear to influence this degradation: the presence of trace levels of moisture (or oxygen?) in the carrier gas, the temperature of the separation, and the length of time the sample spends in the separation process. The use of catalytic driers in the carrier gas line and of higher carrier gas velocities improve the survival rate of such derivatives. Thin-film columns may also help, because the lower partition ratios that accompany these higher phase ratios result in shorter retention times (Section 1.2) under equivalent conditions of temperature and carrier gas velocity.

Changes in the program rate may also have an effect. If the degradation is due to thermal lability, a lower program rate may prove beneficial. If, on the other hand, it is due to total analysis time, a higher rate of temperature programming may improve the survival rate (Section 8.8). By sacrificing some degree of separation efficiency and utilizing wide-bore glass WCOT or SCOT columns, it is possible that better results could be achieved in the analysis of some labile compounds, because such columns result in shorter retention times (Section 1.2).

13.5 Silylation Methods

Because the reagents are highly reactive, they should be stored in a refrigerated dessicator and tested on a known system prior to use. A ten- to fiftyfold excess of silylating agent is usually added to the dry compound being derivatized in a Teflon-lined screw-capped heavy-walled glass reaction vial under anhydrous conditions. Pyridine is most frequently used as the solvent, and some workers regard it as a silylation catalyst. References to specific analytical methods are given in Appendix IV.

13.6 Pyrolysis Gas Chromatography

Materials whose volatilities are too low for gas chromatographic analysis can sometimes be subjected to thermal degradation to produce volatile products whose analysis may provide information about the original sample. The technique of thermal degradation applied in this way is known as pyrolysis, and the resulting chromatogram has been termed a "pyrogram."

Most pyrolysis devices consist of microreactors or filament units. The latter are more common because of their greater simplicity, but they also possess some disadvantages. The sample is usually coated directly on the filament or placed in a small container surrounded by the filament. The assembly is placed in the carrier gas flow stream, and the filament is energized to produce volatile fragments that are then carried to the column for separation to produce, it is hoped, a characteristic pattern. Methods have been described for the characterization of plastics, polymers, and coatings [15, 16], copolymers [17-19], sterols [20], microorganisms [21, 22], and ingredients of foods and drugs [23]. In practice the technique may be plagued with secondary reactions and reproducibility is not always good. The development of reliable methods of silylation has siphoned off much of the original allure, and pyrolysis gas chromatography is now restricted largely to a few special problems such as polymer analysis, although some interesting results continue to be reported. Blackwell [24] utilized pyrolysis gas chromatography to analyze the monomer

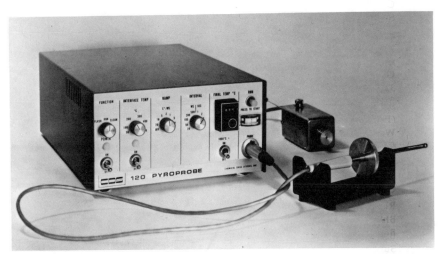

FIGURE 13.2   A commercial pyrolyzer. Because of its close temperature control, some investigators have also used this as a solids injector. (Courtesy of Chemical Data Systems, Inc.)

composition of hexafluoropropylene-vinilidene fluoride copolymers, and Boss and Hazlett [25] used a gold reaction tube in pyrolyzing a number of isomeric alcohols and ketones that were then analyzed by gas chromatography-mass spectrometry. The thermal degradations of aromatic esters have also been explored by pyrolysis GC [26]. Figure 13.2 illustrates a commercial pyrolysis unit. Because of its close temperature control, some workers have also used this at lower temperatures as a "solids injector" with good results.

References

1. Guichon, G., *Chromatographia* **4**, 404 (1971).
2. Schomburg, G., Dielmann, R., Borwitzky, H., and Husmann, H., *J. Chromatogr.* **167**, 337 (1978).
3. Grob, K., and Grob, G., *HRC&CC* **2**, 109 (1979).
4. Madani, C., Chambaz, E. M., Rigaud, M., Durand, J., and Chebroux, P., *J. Chromatogr.* **126**, 161 (1976).
5. Madani, C., and Chambaz, E. M., *Chromatographia* **11**, 725 (1978).
6. Schomburg, G., Hussmann, H., and Weeke, F., *J. Chromatogr.* **99**, 63 (1974).
7. Mathews, R. W., Schwartz, R. D., Pfaffenberger, C. D., Len, S.-I., and Horning, F. C., *J. Chromatogr.* **99**, 51 (1974).
8. *Gas-Chrom News Lett., Appl. Sci.* **16**(2), 1 (1975).
9. Conacher, H. B. S., and Page, B. D., *J. Chromatogr. Sci.* **17**, 188 (1979).
10. Cochrane, W. P., *J. Chromatogr. Sci.* **17**, 124 (1979).
11. Kossa, W. C., and MacGee, J., *J. Chromatogr. Sci.* **17**, 177 (1979).
12. Poole, C. F., and Zlatkis, A., *J. Chromatogr. Sci.* **17**, 114 (1979).
13. Mushau, P., *in* "Handbook of Derivatives for Chromatography" (K. Blay and G. King, eds.), pp. 433-455. Heyden, London, 1978.
14. Sucre, L., Ph.D. Thesis, Univ. of California, Davis. In preparation.
15. Coupe, N. B., Jones, C. E. R., and Stockwell, P. B., *Chromatographia* **6**, 483 (1973).
16. Cucor, P., and Persiani, C., *J. Macromol. Sci., Chem.* **8**, 105 (1974).
17. Okumoto, T., Tsugi, S., Yamamoto, Y., and Takeuchi, T., *Macromolecules* **7**, 376 (1974).
18. Sellier, N., Jones, C. E. R., and Guiochon, G., *J. Chromatogr. Sci.* **13**, 383 (1975).
19. Wallisch, K. L., *J. Appl. Polym. Sci.* **18**, 203 (1974).
20. Gassiot-Matas, M., and Julia-Danes, E., *Chromatographia* **9**, 151 (1976).
21. Meuzelaar, H. L. C., Ficke, H. G., and den Harink, H. C., *J. Chromatogr. Sci.* **13**, 12 (1975).
22. Quinn, P. A., *J. Chromatogr. Sci.* **12**, 796 (1974).
23. Roy, T. A., and Szinai, S. S., *J. Chromatogr. Sci.* **14**, 580 (1976).
24. Blackwell, J. T., *Anal. Chem.* **48**, 1883 (1976).
25. Boss, B. D., and Hazlett, R. N., *Anal. Chem.* **48**, 417 (1976).
26. Sugimura, Y., and Tsuge, S., *J. Chromatogr. Sci.* **17**, 269 (1979).

CHAPTER 14

INSTRUMENT CONVERSION

14.1 General Considerations

Glass capillary columns usually produce superior results when they are installed in equipment specifically designed for their use, but this is not always economically feasible. Although it is sometimes possible to convert older instruments to glass capillary capability, several items demand close attention (and occasionally extensive modification). These include the inlet, the column connections at the inlet and detector, and provision for make-up gas in the instrument itself. The electrometer will prove to be limiting in some instruments. Packed-column peaks are generally broader and are usually at least several seconds wide. Electrometers, recorders, and integrators designed for packed-column use may not possess response (and recovery) times sufficiently short to follow the results of a glass capillary analysis. Provided that the electrometer response is not limiting and that these other shortcomings are recognized, it is often possible to redesign an older instrument so that its performance under these more demanding conditions is, in most cases, satisfactory.

14.2 Electronics

It is probably wisest to begin with a check of the signal amplification and recording capabilities; if these are unsatisfactory, no degree of inlet or detector modification will help. As a first approach, it is useful to verify that response and recovery speeds are adequate. A millivolt source can be used to check the recorder response time, which should be less than 1 (and preferably less than 0.5) sec for full-scale response. A millivolt source or "recorder checker" can be purchased at nominal cost, or a suitable unit can be constructed as shown in Figure 14.1. With the recorder set at zero or baseline, verify that (1) an increase of 1.0 mV achieves full-scale response and (2) the time for the pen to move from 0 to 1 (and from 1 to 0) is within the limits specified when the signal is abruptly applied and abruptly canceled (i.e., switched in single steps rather than swept with the potentiometer).

The speed of the electrometer response can be estimated by subjecting it to a signal, delivered as a short-duration burst, and simulating that generated by an early and well-formed peak. It is unwise to accomplish this by touching an object to the FID elec-

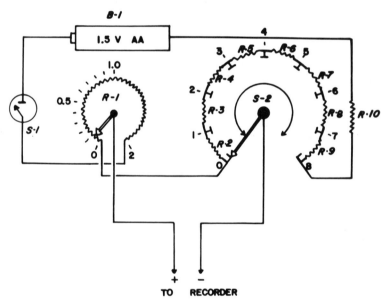

FIGURE 14.1   Schematic of a simple recorder checker. B-1, 1.5-V penlight battery; R-1, 40-ohm linear potentiometer; R-2–R-9, 20-ohm 5% resistors; R-10, 28 kohm resistor; S-1, on–off toggle switch; S-2, ten-position rotary switch. With S-1 "on" the simple voltage divider supplies the sum of the readings at R-1 and S-2 to the recorder test leads.

trode or the disconnected electrometer cable; not only can this approach put a severe jolt on the system, but because the intensity of the signal is unknown, not much useful information is generated. With most chromatographs the "input" or "range" switch affects the signal as it leaves the detector and enters the electrometer, whereas the "output" or "attenuation" switch affects the signal as it leaves the electrometer and enters the recorder. Hence with a proven recorder connected to the chromatograph, some feeling for the speed of the electrometer response can be gained by abruptly shifting the input or range switch. Again, the magnitude of this change may be very large; if this is the case, the response may be immediate and fast when the input is switched to a more sensitive range (i.e., an off-scale signal), but recovery may be slow when it is switched back and the signal eventually decays to a point where the recorder is on scale.

A better method of checking the electrometer response is actually to subject the detector to a very narrow peak. It may be satisfactory to increase the flow rate through the packed column and to try a methane injection. Insertion of a suitable low-dead-volume tee between the column outlet and the detector provides a more positive route. A small injection of methane made through a septum on the side arm of the low-dead-volume tee should provide useful information as to whether the electrometer will be suitable for use with a capillary system. The resultant peak should be narrow and sharp, but it cannot, of course, be of shorter duration than the injection time. Hence it is critical that the injection be executed cleanly and rapidly. Natural gas can normally be utilized as a source of methane. A small, soft-rubber bulb, the type normally used with disposable pipets, can be held in a deflated position and forced on the gas jet. With the gas jet on, the bulb remains inflated with methane, and the sample is obtained by inserting the syringe needle into the bulb. Where methane is not conveniently available, propane or butane is sometimes substituted. Provided the column temperature is not too low or the phase ratio of the column too small (i.e., a thick film; see Figure 15.22), the difference from t_M is usually minor and may not be measurable. Propane is available in small tanks used for propane torches, and butane can be obtained in the same form, or from a disposable cigarette lighter. In this latter case the lever should be triggered slowly so that flame does not ignite; with the butane valve depressed, the syringe needle is inserted into the jet and the plunger retracted. In either case, the injection should be made promptly because of the high diffusivity of these gases. Propane may be an unsatisfactory test substance if there is plastic in the flow stream, because its propensity to adsorb on plastics— including Teflon—may make diagnosis of the peak shape difficult or impossible. Because these gases are already vaporized, the number of moles per unit injection volume is much smaller than with liquefied samples; larger injections are of course necessary and injections of 2–8 μl are normal.

14.3   Inlet Conversion

If machine shop facilities are available, it may be possible to construct an inlet splitter similar to one of the designs shown in Chapter 4. In most cases, however, it is more satisfactory (and more economical) to purchase a commercial kit. Several supply houses offer kits, self-contained or designed to convert a standard ¼-in. heated on-column injector to an inlet splitter, which in some cases also provide splitless and/or on-column injection capability. Particular attention must be directed to unswept volumes, dead-end pockets or blind passages, and areas of excessive volume. Dead volumes should be eliminated, internal volumes should be reduced, and with inlet splitters, the entire inlet should be designed so that it is swept by high-velocity carrier gas. If the carrier gas line is supplied via a flow regulator, this should be bypassed or opened fully, so that the column serves as the only restriction (*exception*: Section 4.4, Figure 4.16). Some manufacturers install a restriction in the carrier gas supply line, usually in the form of a crimped area immediately before the injector. It is wise to ensure that at a reasonable inlet pressure (~1 atm) the carrier gas line is capable of delivering at least 100 cm^3/min as measured at the column inlet connection or the splitter outlet. If this is not the case, the flow may be too restricted for split-mode operation, and the restriction in the carrier gas line should be removed. With some instruments this entails removing the injector, cutting out the restricted portion of the line, and silver-soldering a sleeve (⅛-in. tubing) in position to rejoin the lines. Figures 14.2–14.5 show the components of several commercially available inlet splitter

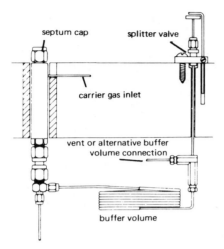

FIGURE 14.2 Schematic of a commercially available inlet splitter kit. (Courtesy of Scientific Glass Engineering, Inc.)

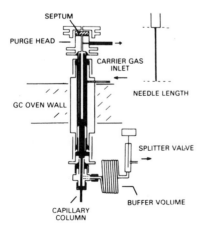

FIGURE 14.3 Flow schematics of a commercially available inlet splitter kit. (Courtesy of Alltech, Inc.)

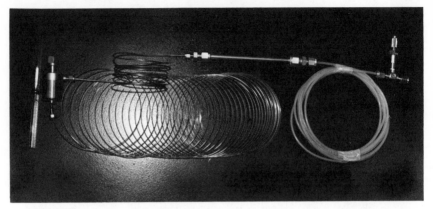

FIGURE 14.4 A commercially available inlet splitter kit designed to accept straightened or unstraightened column ends. (Courtesy of J&W Scientific, Inc.)

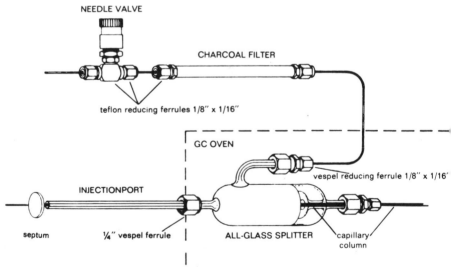

FIGURE 14.5 Schematic of a commercially available inlet splitter kit. (Courtesy of Chrompack, Inc.)

kits. No preference on the part of the author is implied by the selection of these particular models for illustration.

14.4 Make-Up Gas

Optimum gas ratios for the flame ionization detector (FID) are usually about 1 volume of hydrogen to 1 volume of nitrogen or helium carrier gas to 10 volumes of air. Significant departures from the optimum values interfere with combustibility of the mixture and lesser departures affect the sensitivity and response of the detector. The amount of the carrier gas–hydrogen mixture required is dictated by the size of the flame jet: too little fails to maintain the flame; too much blows it out. These interrelated factors have been taken into account in the geometry and design of the electrodes and igniter.

A very few instruments utilize detectors that are designed to operate at the restricted flow rates that characterize capillary columns, but most commercial FID chromatographs are designed for carrier gas flows of 30–60 cm^3/min plus an equivalent amount of hydrogen. With the substitution of a capillary column whose carrier gas flow may be 0.3–3 cm^3/min, sensitivity is drastically decreased unless the 1:1 ratio between carrier gas and hydrogen is maintained, and at these low flow rates, the flame will not endure. The obvious solution is to provide an auxiliary, scavenger, or make-up gas, so that the carrier and auxiliary gases, combined, make up the original 30 cm^3/min, which can then be blended with the original 30 cm^3/min hydrogen. For optimum results and minimum band broadening, it is important that the auxiliary gas be introduced, relative to the column, as shown in Figure 14.6. For the analysis of higher-boiling components, it may be advisable to preheat the make-up gas. Otherwise, condensation of higher-boiling components may occur at this critical point, leading to severe band broadening and noise problems. This defect may also be evidenced by broadened peaks that are studded with positive spikes.

Students are sometimes surprised to find that the addition of a make-up gas—which they quite logically view as dilution—results in increased sensitivity. The answer, of course, lies in the fact that the FID is sensitive to mass–time, not concentration; the

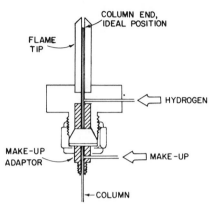

FIGURE 14.6   Schematic of a make-up gas adaptor, for use with straightened column ends. Insertion of the column to this "ideal" position is not always possible; certainly it should terminate above the point of make-up gas introduction, and preferably above both the make-up and hydrogen introduction points. Some manufacturers supply a special flame tip, enlarged at the base to accept the capillary column, for capillary operation.

column delivers the same component mass to the detector as it did before make-up gas was added, but more quickly, and the gas flows are now optimized for maximum detector sensitivity. In a somewhat similar manner, one can also resolve the fears of the investigator who has been utilizing large injections on packed columns to obtain barely detectable peaks, and who is concerned that the small injections to which the WCOT column is limited will result in an overall loss of sensitivity. Because the WCOT column delivers its peaks in very sharp narrow bands, the mass/time ratio is frequently higher, resulting in a more intense signal for a shorter time. This produces tall narrow peaks as opposed to the low broad peaks that result from the larger injection on the packed column.

A suitable make-up gas adaptor can be constructed with minimal machine shop facilities, but again, kits that are available through most supply houses are usually more satisfactory. It may be wise to give some attention to the point of hydrogen introduction also. If this can be introduced together with the make-up gas, the velocity through this critical area will be increased to an even higher value. Figure 14.7 shows a make-up gas adaptor designed for use with unstraightened columns, and Figures 14.8

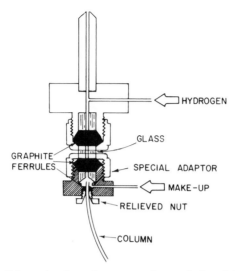

FIGURE 14.7 Schematic of a make-up gas adaptor designed for use with un-straightened capillary columns. The length of the capillary-bore 6-mm glass tube is dependent on oven geometry and the position of the column hanger.

and 14.9 illustrate commercially available make-up gas adaptor kits.

Although hydrogen or helium should normally be used as carrier gas with capillary columns (Sections 8.6, 8.8), the make-up gas should be selected on the basis of detector sensitivity. As a general rule, one should use as make-up whatever gas the manufacturer recommends as a carrier gas when the detector is used with a packed column, e.g., nitrogen for FID and argon–methane for ECD.

14.5   Column Hangers

Hangers suitable for suspending the glass capillary can be obtained commerically (e.g., Figure 14.10), or a crude but satisfactory hanger can be bent from a length of stiff wire, such as a coat hanger. Cages or other protective enclosures appeal to the beginner who is overly concerned about the fragility of the glass capillary, but in most cases these are not only unnecessary but may be undesirable. Most cases of breakage occur at column ends and are of little consequence; in those cases where a column breaks

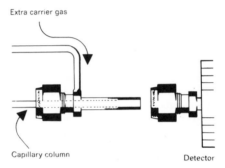

Extra carrier gas

Capillary column

Detector

FIGURE 14.8 Diagram and photograph of a make-up gas adaptor T. (Courtesy of Chrompack, Inc.)

somewhere near the center, one usually ends up with two good columns. If necessary, column segments can be rejoined; heat-shrink Teflon tubing is sometimes used for this purpose, but a fused platinum iridium [1] connection produces superior results. Because of their flexibility and superior strength, these problems rarely occur with fused silica columns. If the cage or enclosure contributes thermal mass or restricts the flow of oven air around the column, it can be a definite disadvantage. The ability of the column to respond evenly and uniformly to oven temperature changes is then inhibited, and this can lead to broader ranges of partition ratios for each solute. Columns are occasionally designed to lie on a horizontal rack; for these same reasons such racks should be of light-gauge metal and largely open, similar to a coarse screen.

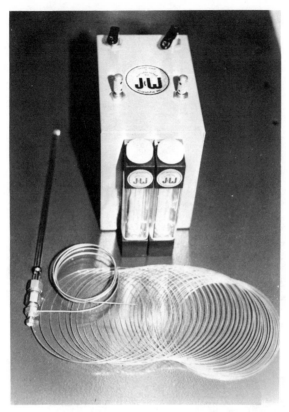

FIGURE 14.9   Make-up gas adaptor kit with flow control module for make-up and hydrogen gases, designed to accept straightened or unstraightened column ends. (Courtesy of J&W Scientific, Inc.)

14.6   Gas Supplies

Gases are preferably supplied as shown schematically in Figure 14.11; a separate regulator, supplied via a toggle valve, should be used for each gas demand function of each instrument; i.e. four such units per instrument are desirable with FID. Once the gas flows are optimized (*vide infra*), the toggle valves are used to energize or shut down each particular gas supply. Molecular sieve cartridges should be used on each gas line as it leaves the supply tank. These can be constructed of $\frac{1}{4}$-in. pipe and should be pe-

FIGURE 14.10 A simple column hanger, designed for attachment to the roof of the column oven with sheet metal screws. Three to four coils of column should be left free and unsupported at either end to provide flexibility at the points of column attachment.

riodically regenerated by heating to 300°C under a reverse flow of clean nitrogen (passed through freshly regenerated molecular sieve). The carrier gas line should be equipped with a particle filter (dust guard) and, depending on the liquid phase (Section 10.2), drier and/or oxygen scrubber cartridges.

Most pressure regulators operate more satisfactorily through a flow restrictor; the column serves this function in the carrier gas

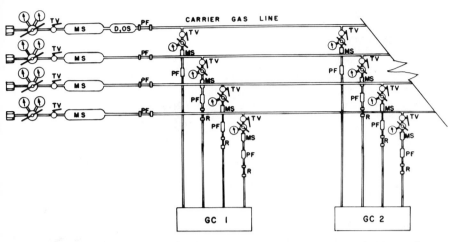

FIGURE 14.11 Schematic of a preferred method for gas supply. Main supply lines are fed by two-stage regulators, followed by toggle valves. There is usually sufficient buffer volume in the line that by closing this toggle valve, the regulator can be transferred to another cylinder without disturbing a chromatographic run. Molecular sieve filters (MS, 13A plus 4A or 5A) should be used on each line, and these should be followed by particle filters (PF) and restrictors (R); the restrictor is usually not required for the carrier gas line. Depending on the liquid phase used, driers (D) and/or oxygen scrubbers (OS) may be desirable on the carrier gas line. A single-stage regulator supplies each gas demand function of each instrument via a toggle valve.

line, but the restriction offered by the FID is usually inadequate, and an in-line restrictor is advantageous. Pretorius [2] described a simple method for twisting a short length of flattened tubing to achieve the necessary restriction (Figure 14.12).

14.7 Checkout

With the carrier gas off and the oven at room temperature, the hydrogen is energized and adjusted to the manufacturer's recommendation for that FID (normally ~30 cm³/min). The carrier and make-up gases are then turned on, and the latter is adjusted to deliver a flow equivalent to the manufacturer's recommendation for carrier gas flow (normally ~30 cm³/min). (The make-up gas flow can be adjusted with the hydrogen on if the two rec-

FIGURE 14.12   Simple but effective flow restrictor. The tube, ⅛-in. copper or stainless steel (heavy wall preferred), is flattened in a machinist's vise and twisted to achieve the desired degree of restriction, thus permitting the regulator to function properly. (After Pretorius [2].)

ommendations are added.) Air is then supplied (the recommended flow is usually in the range of 200–300 cm³/min) and the flame lighted. Gases may have to remain on for 2 or 3 min before the flame will light.

It is advisable to use a short—5–10-m—column for the initial checkout. Not only are results obtained more quickly, but inadequacies in the operator's injection technique are more evident. It can be a humbling experience for the experienced packed-column chromatographer to find that injection techniques he has used with good results for many years produce, in this more demanding system, doublets or triplets for each peak. No longer can he use an unwiped needle, insert a syringe with a sample-filled needle, or make delayed or slow injections without paying a severe penalty. Figure 14.13 shows what can happen when slow, wet, or hot needle (Section 4.8) injections are used with split injection. The defect—doublet or multiple peaks for each solute—is more severe with shorter columns and early peaks. A similar phenomenon can occur when a segment of broken column has been allowed to remain in the splitter flow path (Section 4.8). In this latter case, the doublet peaks persist into higher-k solutes.

Initial injections should utilize methane, and the average linear

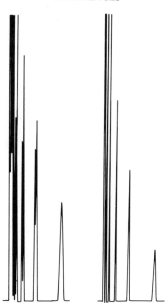

FIGURE 14.13 Results of faulty injection technique with split injection. Left, a simple six-component mixture, 0.2 μl split ~1:100, "hot needle" injection (Section 4.7). Right, same mixture, rapidly injected through emptied needle that had been carefully wiped. Note that multiplicity is most noticeable with early peaks, and that the degree of differentiation decreases with peaks of longer retention; with the final component, multiplicity is no longer apparent. The last component is skewed, due to overloading.

carrier gas velocity should be adjusted to ~25 cm/sec for helium carrier gas or 40 cm/sec for hydrogen carrier gas. Ensure that the split ratio is in the proper range. Solutions suitable for these initial tests have been described in Section 5.4. If peaks are sharp and well formed, the instrument is ready for use; if difficulties are encountered, the problem should be diagnosed (Chapter 16) and attention directed to the points detailed in Section 14.2.

References

1. Jennings, W., and Shibamoto, T., "Qualitative Analysis of Flavor and Fragrance Volatiles with Glass Capillary Columns," Academic Press, New York, 1979. In press.
2. Pretorius, V., HRC&CC 2, 186 (1979).

CHAPTER 15

APPLICATIONS OF GLASS CAPILLARY GAS CHROMATOGRAPHY

15.1 General Considerations

Gas chromatography continues to enjoy great popularity because it is the best available method for the separation of volatile mixtures, analysis times are reasonably short, and the method usually exhibits good sensitivity. Although the bulk of gas chromatographic applications have employed packed columns, results from those studies that have utilized glass capillary columns make it inevitable that the use of the latter will continue to grow, largely at the expense of the former. Glass capillary gas chromatography possesses to a higher degree all the advantages of packed-column gas chromatography and offers some additional inducements of its own.

Resolution (the ability to separate the components of a mixture) is greatly enhanced by the glass capillary. Novotny *et al.* [1] emphasized that because of the tremendous complexity of volatile samples derived from food aroma collection, air pollution analysis, tobacco smoke, and physiological fluids, high-efficiency capillary columns are required to obtain an adequate degree of resolution. This much higher level of resolution permits the analyst to obtain a superior separation in the same time required for a packed-column analysis, or to obtain a separation equivalent to that of the packed column in a fraction of that time; in other words, analysis times can be made much shorter [2, 3].

Although the total amount of a given compound delivered to the detector is very much less with the glass capillary column, the peak is so much sharper that the concentration per unit time is usually higher; hence sensitivity is normally enhanced by the glass capillary column as compared to the packed column. Additionally, many of the components of such systems (as well as some pesticide chemicals and derivatized drugs, pharmaceuticals, amino acids, steroids, and saccharides) suffer severe attrition or entirely fail to reach the detector when subjected to analysis in conventional gas chromatographic equipment. Consequently, there has been a range of activity on the part of individual investigators in the application of the more inert and highly efficient glass open tubular columns to specific types of analyses. Some of these efforts have involved detailed studies of a given class of compound, and others represent a cursory analysis intended to demonstrate to other investigators the advantages of these systems.

This section is intended more as a representative sampling rather than as an exhaustive survey of efforts reporting the application of analyses in glass open tubular columns to various fields of endeavor. It should be noted that the chromatography in some of these examples is capable of improvement. Defects that could be cited include plumbing problems, as evidenced by tailing of all peaks, including hydrocarbons, overloading, loss of deactivation, poor choice of carrier gas and/or carrier gas velocity, and poor choice of temperature and/or program conditions. In some cases the defect was inadvertent, but useful data—far superior to those attainable with packed columns—were generated; in other cases resolution may have been deliberately sacrificed in favor of speed or vice versa.

15.2   Air, Smoke, and PAH Analyses

Some 28 polyclyclic and polycylic aromatic hydrocarbons, including seven sulfur-containing compounds, were found in carbon black from sulfur-containing petroleum feedstocks by analysis on a short WCOT glass column coated with SE 52 [4]. When the analysis was repeated on a packed stainless steel column, not only was a much lower degree of separation obtained, but the sulfur-containing compounds were no longer apparent. The glass capillary achieved a much higher degree of resolution, permitting the analysis of previously unresolved isomers and trace compounds.

Lee *et al.* [5] used a combination of liquid chromatography fractionations to prepare selectively enriched extracts of tobacco and marijuana smoke condensates, which were then analyzed by glass capillary gas chromatography. Novotny *et al.* [6] used the porous polymer Tenax GC to concentrate smoke from three different types of cigarettes. The concentrates were desorbed and chromatographed on glass WCOT columns. (See Figure 15.1.) In another study, over 100 polycyclics, including trace alkylated compounds, were separated from air pollution particulate matter. Superb resolution, extending to isomeric compounds that differed in alkyl group position, was demonstrated [7].

Rapp *et al.* [8] coupled a glass capillary column directly to a mass spectrometer for the identification of separated components from a tobacco smoke condensate. They achieved wide-spaced separation of phenanthrene, anthracene, fluoranthene, pyrene, chrysene, perylene, benzypyrene, coronene, and the $C_{14,16,18,20,22,24,28,32,34}$ hydrocarbons. Also separated was a test mixture consisting of limonene, furfuryl alcohol, nicotine, and phenol.

Onuska and Comba used a glass capillary with an extensively whiskered inner surface on a mixture of polynuclear aromatic hydrocarbons [9] (Figure 15.2) and Jenkins [10] optimized conditions on a 15-m column to achieve excellent separation of a standard mixture of polynuclear aromatic hydrocarbons (Figure 15.3).

Roeraade [11] used a 0.2-mm × 9.6-m WCOT glass capillary on cigarette smoke (Figure 15.4), and Grob [12] achieved superb resolution of a tobacco smoke fraction on a 0.32-mm × 145-m WCOT glass capillary coated with Emulphor ON 870.

In a study of the carbonyl compounds present in tobacco smoke and automotive exhaust, Hoshika and Takata [13] prepared the 2,4-dinitrophenylhydrazones, which were then separated on a 0.25-mm × 20-m glass capillary coated with SF 96. Except for the derivatives of *n*-valeraldehyde and isobutylmethyl ketone, whose

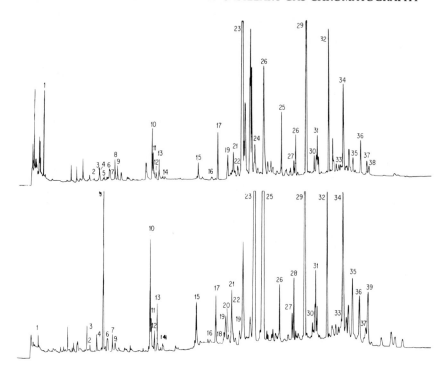

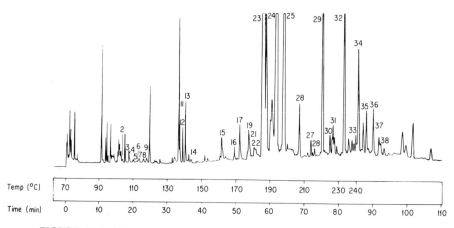

FIGURE 15.1   Chromatograms of extracts of marijuana from three different sources. 11-m × 0.26-mm glass capillary column coated with SE 52 [6].

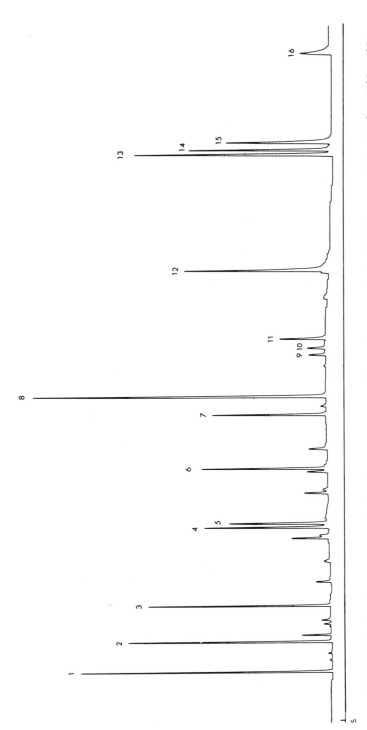

FIGURE 15.2 Polynuclear aromatic hydrocarbons on a glass capillary column coated with OV 3. Programmed from 60° to 230°C at 2°C/min. 1, biphenyl; 2, acenaphthalene; 3, fluorene; 4, phenanthrene; 5, anthracene; 6, 9-methylphenanthrene; 7, fluoranthrene; 8, pyrene; 9, benzo(a)fluorene; 10, benzo(b)fluorene; 11, 1-methylpyrene; 12, triphenylene; 13, benzo(e)pyrene; 14, benzo(a)pyrene; 15, perylene; 16, dibenz(a,c)anthracene. Whisker column. (From Onuska and Comba [9].)

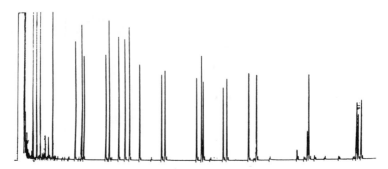

FIGURE 15.3   PAH standard mixture on a 15-m × 0.24-mm glass capillary coated with SE 54 and programmed from 30° to 280°C at 4°C/min. Hydrogen carrier at $\bar{u}_{\text{initial}}$ = 62.5 cm/sec and $\bar{u}_{\text{final}}$ = 40 cm/sec. Injection, 2 μl splitless. Mixture includes hexamethylbenzene;   o-xylene;   1-propylbenzene;   n-propylbenzene; indan; 1,2,3,4-tetramethylbenzene; naphthalene; benzothiophene; 2-methylnaphthalene; 1-methylnaphthalene; 1,3,5-triisopropylbenzene; biphenyl; 2,6-dimethylnaphthalene; 2,3,5-trimethylnaphthalene; fluorene; dibenzothiophene; phenanthrene; anthracene; dodecylbenzene; 1-methylphenanthrene; fluoranthrene; pyrene; benz(a)anthracene; chrysene; benz(e)pyrene; benz(a)pyrene; perylene. (From Jenkins [10].)

peaks overlapped, and the o- and m-tolvaldehyde derivatives, which were poorly separated, they reported complete separation of the 2,4-DNPH derivatives of ten aliphatic aldehydes, eight aliphatic ketones, and four aromatic aldehydes.

A specific method for fingerprinting marijuana samples has been reported [14]. The sample is subjected to Soxhlet extraction

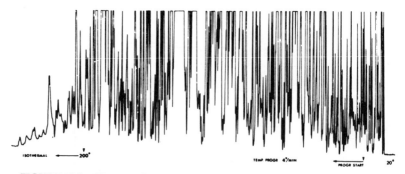

FIGURE 15.4   Glass capillary chromatogram of a tobacco smoke fraction. (From Roeraade [11]; reprinted by permission of the copyright owner.)

with cyclohexane, and the extracts washed with nitromethane. The washed cyclohexane extracts were evaporated to dryness under vacuum, redissolved in dichloromethane, and concentrated on a short precolumn. The volatiles were transferred to and analyzed on an 11-m × 0.26-mm WCOT glass capillary column coated with SE 52. Chromatograms were reported to be reproducible and major differences were demonstrated between Turkish and Mexican marijuana. Differences were also apparent between chromotograms of Mexican-grown and Indiana-grown Mexican marijuana.

Figure 15.5 shows chromatograms of polynuclear aromatic hydrocarbons found on urban air particulates and analyzed by simultaneous ECD–FID [15]. Bertsch [16] has authored a review of the application of glass capillary GC to air pollutant analysis.

15.3 Amino Acid Analysis

The suitability of glass capillary gas chromatography to the separation of amino acid derivatives has been explored by several workers. Deyl [17] reported that while the phenylthiohydantoin derivatives offered a strong potential, the trimethylsilyl derivatives were also very useful. Chromatography of some acetylated phenylthiohydantoin derivatives that failed to survive analysis in stainless steel columns was also demonstrated. He also achieved reasonable separation of the methyl esters of the N-pivalylamino acids on a glass capillary column that had been coated with a mixture of XE 60 and FFAP.

Eyem and Sjøquist [18] used a short fine-bore glass capillary coated with a mixture of OV 101 and OV 225 to achieve the separation of silylated methyl thiohydantoin derivatives of amino acids. Provided the cysteinyl and arginyl residues had been converted to S-methylcysteinyl and ornithyl derivatives, respectively, 20 amino acids could be separated in a single run. Some of the derivatives showed a typical double-peak pattern that the authors found to be of diagnostic value. Isoleucine, for example, produced a doublet, which was ascribed to the formation of MTH-allo-isoleucine during the silylation reaction. Although slow program rates achieved superior separation, partial decomposition of certain amino acid derivatives was observed, and the histidine derivative occasionally disappeared entirely. Faster program rates

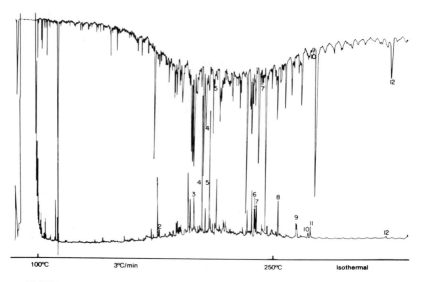

100°C   3°C/min   250°C   Isothermal

FIGURE 15.5 Polynuclear aromatic hydrocarbons on urban air particulates. Sample trapped on glass fiber filter and analyzed on a 50-m × 0.35-mm glass capillary coated with SE 54, programmed from 100° to 250°C at 3°C/min. Splitless injection. Top, ECD; bottom, FID [15].

Peak no.	Identity	Peak no.	Identity
1	2-methyl naphthalene	7	dibenzofuran
2	1-methyl naphthalene	8	fluorene
3	biphenyl	9	9-methylfluorene
4	acenaphthylene	10	9, 10-dihydroanthracene
5	acenaphthene	11	2-methylfluorene
6	4-methylbiphenyl	12	1-methylfluorene

reduced the degree of separation, but less decomposition occurred.

Cavadore et al. [19] used an FFAP-coated glass capillary column for the separation of the methylesters of benzoyl and pivalyl amino acid derivatives, and Schomburg and Husmann [20] demonstrated the separation of racemic amino acid esters on a specially prepared liquid phase.

The use of a nitrogen-selective detector with glass capillary columns was explored by Adams et al. [21]. Amino acids from model systems and biological samples were isolated by ion exchange and chromatographed as *n*-propyl, *N*-acetyl derivatives. Poole and Verzele examined the separation of *N* (*O*) acyl alkyl ester amino acid derivatives on SE 30, OV 210, OV 17, and EGA coated columns; the latter required a whisker column (Section 2.2) and extensive deactivation [22].

Various enantiomeric esters of TFA-α-amino acids were separated on a column coated with an optically active liquid phase by Abe *et al.* [23]. The column, coated with *N*-caproyl-L-valyl-L-valine cyclohexyl ester gave clean separations of alanine, α-amino-*n*-butyric acid, valine, threonine, alloisoleucine, isoleucine, norvaline, leucine, proline, norleucine, serine, cysteine, aspartic acid, methionine, phenylalanine, and glutamic acid. The use of chiral stationary phases for amino acid enantiomer separation has also been explored by Nicholson *et al.* [24].

The quantitative determination of amino acids has been one of the more difficult gas chromatographic determinations and one of the few where some advantages have been claimed for packed columns. Some of the amino acid derivatives are thermally labile and suffer severe loss during the volatilization step. On-column injections, under which the sample is exposed to the lowest possible temperatures, have in the past been very difficult to accomplish with small-bore capillary systems. Because this is a straightforward technique with packed columns, many analysts have elected to forego the increased speed, resolution, and sensitivity offered by the capillary system in favor of a less destructive sample introduction. The capillary-compatible on-column injection system described by Grob and Grob [25, 26] and discussed in Section 4.5 may well resolve this problem. The application of glass capillary columns to amino acid analysis has been reviewed by Jaeger *et al.* [27].

15.4 Drugs and Pharmaceuticals

Rijks and Cramers [28] presented data showing the separation of a mixture of barbiturates on a glass micropacked column containing a mixture of OV 17 and OV 225 on Gaschrom Q. Included in the cleanly separated test mixture were aprobarbital, amobarbital, secobarbital, hexobarbital, brallobarbital, heptobarbital, and heptabarbital. These same authors used a second micropacked column coated with OV 101 and potassium hydroxide.

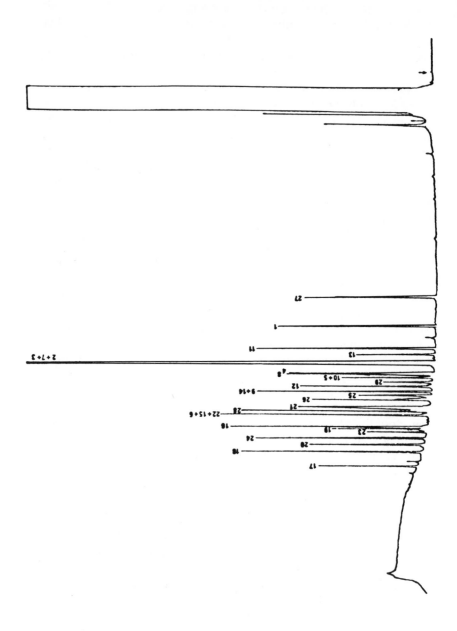

The test mixture, which again was well resolved, consisted of methyl, ethyl, isopropyl, methyl-ethyl, butyl, and methyl-isopropyl amphetamine.

A mixture consisting of cyclopentamine, propylhexedrine, methamphetamine, amphetamine, mephentermine, phendimetrazine, ephedrine, phenmetrazine, phenylpropanolamine, and benzphetamine was separated on a glass WCOT column by Schomburg and Husmann in one instance, and a mixture of phendimetrazine, ephedrine, phenmetrazine, phenylpropanolamine, benzphetamine, xylometazoline, tetrahydrozoline, caffeine, cocaine, naphthazoline, and oxymetazoline in another [20]. Dünges et al. [29] described a method based on multiple reactions to produce several different derivatives from micro samples. Results of these analyses were suitable for barbiturate classification. A chromatogram of the products of one such derivatization is shown in Figure 15.6.

15.5 Essential Oil Analysis

Man's interest in the essential oils can be traced as far back as our earliest historical records. Essential oils and aromatic sub-

←

FIGURE 15.6 Splitless injection of 2-μl reaction mixture with allyl bromide. Column, 40 m × 0.3 mm, SF 96. Programmed from 50°C at 4°C/min. This represents one of a series of reactions to produce derivatives from which barbiturate classification was possible. The example isolates those containing an allyl or similar group in position 5. (From Dünges et al. [29].)

Peak no.	Identity	Peak no.	Identity
6	aprobarbital, 5-allyl-5 isopropylbarbituric acid	10	nealbital, 5-allyl-5-neo-pentylbarbituric acid
7	butalbital, 5-allyl-5-isobutyl barbituric acid	11	allobarbital, 5-5-diallyl barbituric acid
8	talbutal, 5-allyl-5-(sec-butyl) barbituric acid	12	butylvinyl, 5-(1 methyl butyl)-5-vinyl barbituric acid
9	secobarbital, 5-allyl-5(1-methylbutyl) barbituric acid)	13	crotarbital, 5-crotonyl barbituric acid
		14	vinbarbital, 5-ethyl-5 (1-methyl-buten-1-yl) barbituric acid

stances were major items of commerce and were instrumental in the establishment of early trade routes. Most are relatively complex mixtures, and prefractionation may be desirable (Section 12.9).

Figure 15.7 shows three chromatograms of the essential oil of peppermint, which also serve to illustrate our increased sophistication in gas chromatographic analysis. Even so, groupings of the monoterpene hydrocarbons and sesquiterpene hydrocarbons complicate the glass capillary separation and indicate the desirability of multiramp programming or prefractionations. Figure 15.8 shows the essential oil of a juniper leaf, on three different liquid phases. None of these alone achieved a complete separation of this complex mixture. Some of the sesquiterpene hydrocarbons (peaks 56–88) are poorly resolved on the Carbowax 20 M column, while the methyl silicone column gave a less satisfactory separation of the more volatile components. Figure 15.9 illustrates a separation of a steam-distilled Bulgarian rose oil, and a commercial perfume, Rive Gauche, is shown in Figure 15.10. Chromatograms derived by collecting the volatile emanations from two varieties of rose are shown in Figure 15.11 [30].

15.6   Fatty Acid Analysis

The analysis of fatty acid mixtures, exhibiting a range of unsaturation and isomeric forms, is particularly well suited to the extremely high resolution that can be achieved with glass capillary columns. Schomburg and Husmann [20] studied the effect of several parameters on the separation of the C_{10} through C_{18} fatty acids, and the C_{18} through C_{26} fatty acid methyl esters on glass capillaries coated with Carbowax 20 M and with DEGS. Badings et al. [31] showed the separation of the methyl esters of the fatty acids of milk fat, and a particularly striking separation of the methyl esters of the whole range of C_{10}–C_{24} fatty acids was presented by Jaeger et al. [32] (Figure 15.12). Excellent separations of complex fatty acid mixtures have also been achieved by Onuska and Comba [9]. The separation of isomeric mixtures of the mono-unsaturated C_{18} fatty acid methyl esters was explored by van Vleet and Quinn [33], who used a preliminary HPLC separation to yield LC fractions that were then analyzed on glass capillary columns coated with a cyano silicone SP 2340. Grob et al. [34] also

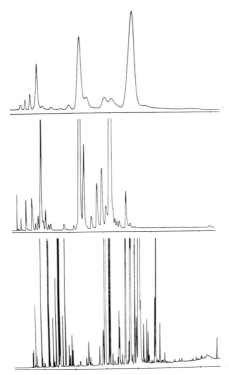

FIGURE 15.7 Chromatograms of oil of peppermint. Top, a ¼-in. × 6-ft packed column, middle, 500-ft × 0.03-in. stainless steel capillary, and bottom, an 80-m × 0.25-mm glass capillary column. Methyl silicone liquid phase in each case.

examined the application of high-polarity liquid phases to these separations. While the relative retentions of the isomers were higher in these more polar phases, column efficiencies were, as one would expect (Section 11.1), generally lower as reflected by the apparent theoretical plates obtained. Recycle chromatography (Section 9.4) may offer the best approach to these difficult separations. The application of glass capillary gas chromatography to fatty acid separations is the subject of a current review [35].

15.7 Food and Beverage Analysis

There has been a great deal of interest in meat flavors, which is generally related to the development of artificial flavors that

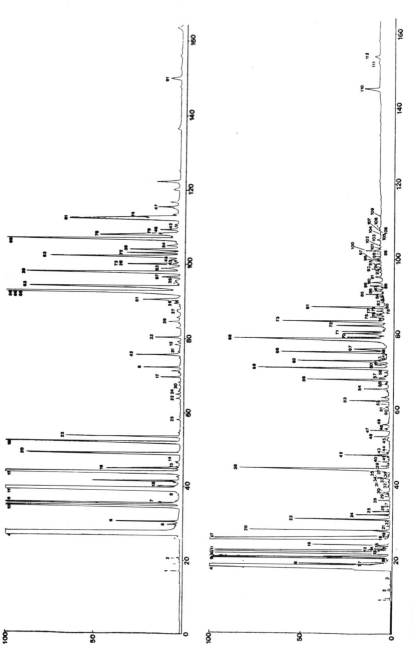

FIGURE 15.8 Leaf oil of a juniperlike shrub on glass capillaries coated with Carbowax 20 M (top), FFAP (center), and SE 30 (bottom)

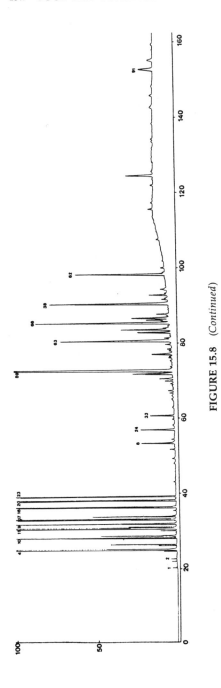

FIGURE 15.8 (Continued)

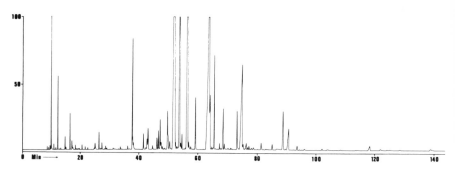

FIGURE 15.9   Chromatogram of a steam-distilled Bulgarian rose oil on a 50-m × 0.25-mm column coated with Carbowax 20 M and programmed from 80° to 200°C at 2°C/min; split injection.

might be utilized with unconventional foods and help relieve the world food problem. In spite of this activity, attempts to reconstitute acceptable meat flavors based on product analyses have met with limited success. In a recent study, Shibamoto and Russell [36] utilized a model system composed of D-glucose, hydrogen sulfide, and ammonia, which was heated to 100°C for 2 hr. Following steam distillation the reaction products were extracted with dichloromethane and subjected to gas chromatographic analysis in an all-glass system, utilizing both Carbowax 20 M and SE 30 WCOT glass capillary columns. (See Figures 15.13 and 15.14.) Identifications were based on integrated gas chromatography–mass spectrometry (GC–MS) in which glass capillary columns

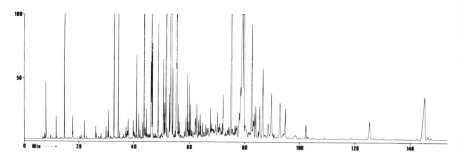

FIGURE 15.10   Chromatogram of a Rive Gauche perfume on Carbowax 20 M. Conditions same as in Figure 15.9.

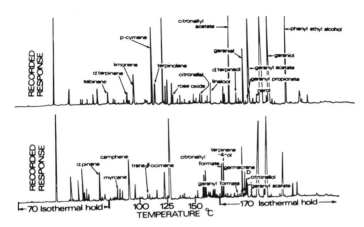

FIGURE 15.11 Glass capillary chromatograms of headspace volatiles from two varieties of rose. (From Sirikulvadnana *et al.* [30]; reprinted by permission of the copyright owner.)

were directly interfaced with a quadrapole-type mass spectrometer. A wide range of aliphatic and cyclic products was isolated, including sulfur- and nitrogen-containing compounds. The volatiles included sulfides, thiols, thiophenes, thiazoles, pyrazines, and furan derivatives. Several of the isolated components were reported to possess cooked-meat aromas.

In a similar analysis applied to canned pork meat subject to different heating treatments [37], sulfur compounds were again isolated, as well as a number of aliphatic alcohols.

The flavor compounds of a number of fruits, including peaches [38] and figs [39], have been analyzed on glass capillary columns. The netted orange-fleshed variety of muskmelon, known to most of the United States as cantaloupe, was studied in terms of varietal differences [40] (Figure 15.15), ripening patterns [41], and the route of biosynthesis of the flavor compounds [42]. Badings *et al.* showed excellent separation of volatile aroma mixtures (sources unidentified) composed largely of aldehydes, esters, and ketones on glass capillaries coated with SE 30, and carbonized columns coated with UCON HB 5100 [31].

Beer volatiles [43] have been isolated by sweeping the products with nitrogen and trapping the entrained volatiles on porous

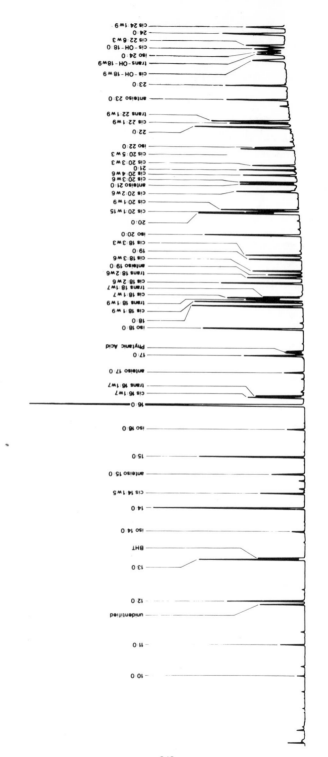

FIGURE 15.12 Glass capillary chromatogram of a standard mixture of fatty acid methyl esters. 50-m column coated with FFAP, and programmed from 95° to 198°C at 1°C/min. Hydrogen carrier gas. (From Jaeger et al. [32]; reprinted by permission of the copyright owner.)

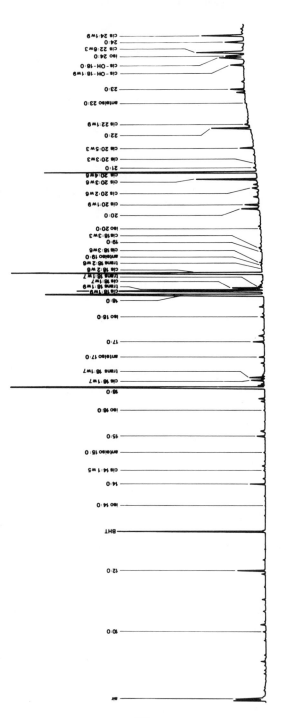

FIGURE 15.12 (Continued)

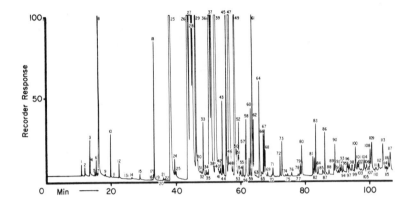

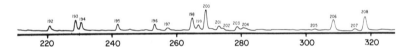

FIGURE 15.13   Volatiles from a meat flavor model system. Reaction products from ammonia–glucose on a 100-m × 0.25-mm glass capillary column coated with Carbowax 20 M. Programmed from 70° to 170°C at 1°C/min. (Courtesy of Dr. T. Shibamoto, Ogawa and Company, Tokyo.)

polymers (Section 12.10). The recovered essences were then sub-jected to gas chromatographic analysis on glass capillary systems.

Glass WCOT columns have also been utilized in studies on the flavor essences of wines and fruits [38], and many essential oils (e.g., [44, 45]). Klimes and Lamparsky [46] applied headspace techniques to a study of the volatile emanations of vanilla beans (Figure 15.16), and Kolb *et al.* [47] used an automated headspace sampling apparatus to produce headspace chromatograms of cof-fee aroma, trapped by charcoal adsorption and eluted with benzyl alcohol (Figure 15.17).

Gas chromatography is also widely used in quality control op-

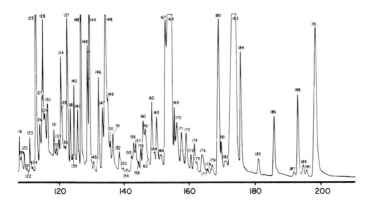

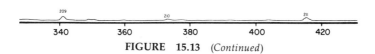

FIGURE 15.13 (*Continued*)

erations involving foodstuffs. Bjørseth (e.g., [48]) applied glass capillar gas chromatography to the identification of polynuclear aromatic hydrocarbons in mussels from Norwegian fjords, and Meili *et al.* [49] used a photoionization detector with glass capillary columns to determine nitrosamines in several meat products.

15.8 Hydrocarbon and Coal Applications

The petroleum chemist played a critically important role in the development of capillary gas chromatography. However, because he was normally concerned with hydrocarbon analysis, adsorption was rarely a problem, and with a few notable exceptions, most efforts utilized metal capillary columns. Several workers have now demonstrated that the higher coating efficiencies that

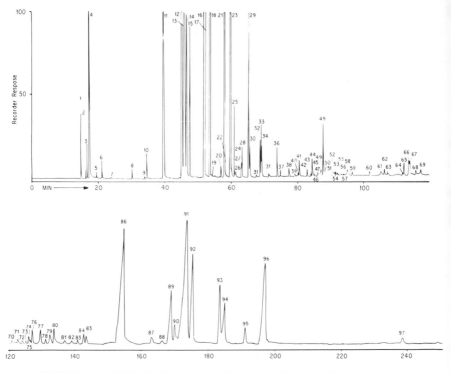

FIGURE 15.14   Volatiles from a meat flavor model system. Reaction products of hydrogen sulfide, ammonia, and glucose. 100 m × 0.25 mm glass capillary column coated with Carbowax 20 M. Programmed from 70° to 170°C at 1°C/min. (Courtesy of Dr. T. Shibamoto, Ogawa and Company, Tokyo.)

are obtainable with glass capillary columns not only provide more complete separations of these very complex mixtures but also permit the analysis of reactive (e.g., sulfur- and phenol-containing) components. Figure 15.18 shows a chromatogram of one restricted fraction of a crude oil [50]. The complexity of a normal gasoline is demonstrated in Figure 15.19 [51], and Figure 15.20 demonstrates that by sacrificing a large degree of resolution (i.e., short columns, high flow rates, fast programs), rapid screening analyses can be performed in a small fraction of the normal analysis time [3]. Using a 20-m column programmed to 330°C, and mass spectrometric detection, Dielmann et al. [52] were able to chromatograph hydrocarbons as large as C_{57} (Figure 15.21).

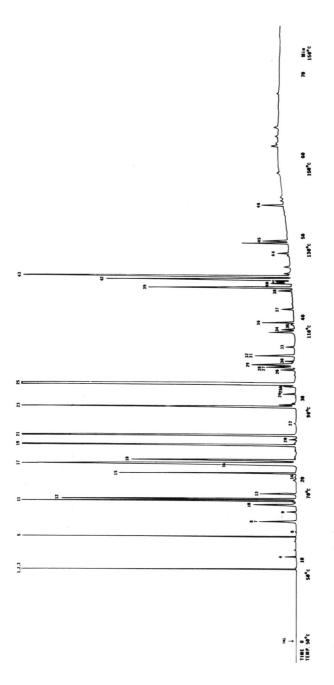

FIGURE 15.15 Volatile components of a muskmelon on an 80-m × 0.25-mm glass capillary coated with Carbowax 20 M. (From Yabumoto et al. [40]; reprinted by permission of the copyright owner.)

247

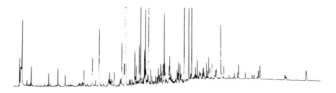

FIGURE 15.16   Chromatogram of vanilla bean headspace, trapped on charcoal and eluted with carbon disulfide. (From Klimes and Lamparsky [46].) Column, 50 m × 0.31 mm coated with UCON HB 5100 programmed from 20° to 180°C at 3°C/min. Splitless injection of 8 μl sample.

A thick-film column ($d_f \simeq 0.9\ \mu$) was used by Johansen [53] in the separation of low-molecular-weight hydrocarbons (Figure 15.22) and chlorinated hydrocarbons (Figure 15.23). Although the number of theoretical plates is lower [because the thick film increases the contribution of C_L to h (Section 8.6)], component

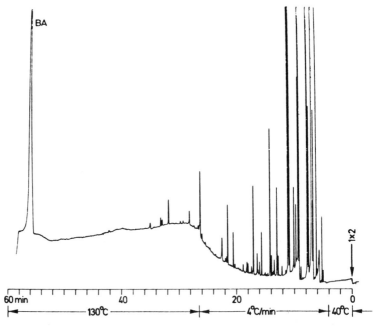

FIGURE 15.17   Headspace volatiles of coffee, trapped on charcoal and eluted with benzyl alcohol. Column, 43 m × 0.2 mm coated with UCON LB 550; 5-sec injection from pressurized headspace vial programmed as shown. (From Kolb *et al.* [47].)

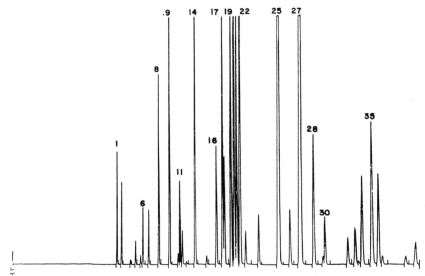

FIGURE 15.18 Chromatogram of 85°–114°C fraction of an East Texas crude oil on a 300-ft × 0.01-in. (~91-m × .25-mm) column coated with a hexadecane–Kel F mixture; isothermal at room temperature [50].

Peak no.	Identity	Peak no.	Identity
1	isopentane	2	n-pentane
3	2,2-dimethylbenzene	4	cyclopentane
5	2,3-dimethylbutane	6	2-methylpentane
7	3-methylpentane	8	n-hexane
9	methylcyclopentane	10	2,2-dimethylpentane
11	benzene	12	2,4-dimethylpentane
13	2,2,3-trimethylbutane	14	cyclohexane
15	3,3-dimethylpentane	16	1,1-dimethylcyclopentane
17	2 methylhexane	18	2,3-dimethylpentane
19	1-*cis*-3-dimethylcyclopentene	20	3-methylhexane
21	1-*trans*-3-dimethylcyclopentene	22	1-*trans*-2-dimethylcyclopentene
23	3-ethylpentane	24	2,2,4-trimethylpentane
25	n-heptane	26	1-*cis*-2-dimethylcyclopentene
27	methylcyclohexane	28	1,1,3-trimethylcyclopentane
29	2,2-dimethylhexane	30	ethylcyclopentane
31	2,5-dimethylhexane	32	2,4-dimethylhexane
33	2,2,3-trimethylpentane	34	1-*trans*-2-*cis*-4-trimethylcyclopentene
35	toluene	36	1-*trans*-2-*cis*-3-trimethylcyclopentene
37	3,3-dimethylhexane	38	2,3,4-trimethylpentane
39	1,1,2-trimethylcyclopentane		

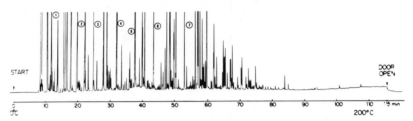

FIGURE 15.19   Regular gasoline. Column, 100 m × 0.25 mm, coated with SF 96. Initial hold 8 min 0; 2°C/min to 200°C; split injection. Circled numbers: 1, *n*-pentane; 2, *n*-hexane; 3, benzene; 4, *n*-heptane; 5, toluene; 6, *n*-octane; 7, *n*-nonane. (From Averill and March [51].)

separation—and the retention of low boilers—is enhanced to such a degree that base line separation of methane and ethane have been achieved at a column temperature of 69°C. Columns of more conventional film thickness (d_f = ~0.4 nm) require cooling to approximately 20°C to achieve this separation.

Yang and Cram [54] used glass capillary GC and selective detection to study sulfur- (Figure 15.24) and phosphorus-containing

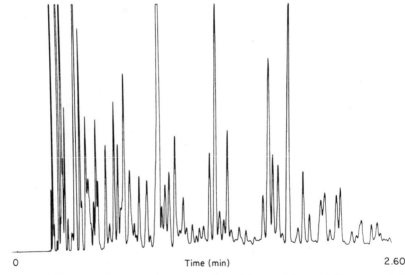

FIGURE 15.20   Fast screening of gasoline. Column, 10 m × 0.25 mm, coated with SP 2100, programmed from 30° to 200°C at 35°C/min. (From Rooney *et al.* [3].)

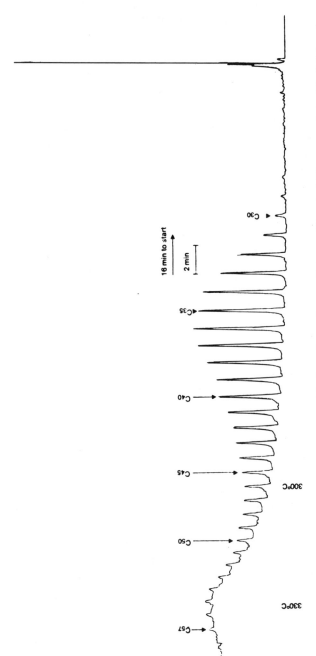

FIGURE 15.21 Total ion current chromatogram of hydrocarbon mixture. Column, 20 m × 0.25 mm, coated with OV 1 and programmed from 180° to 330°C at 4°C/min. (From Dielmann et al. [52].)

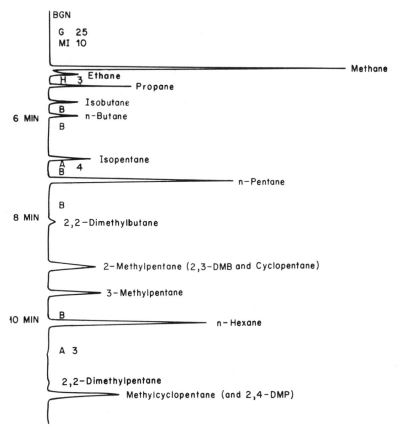

FIGURE 15.22   Low-molecular-weight hydrocarbons on a 55-m × 0.27-mm column with thick film of OV 101 liquid phase; $d_f = 0.9 \mu$. Temperature, 69°C isothermal, split injection [53].

(Figure 15.25) compounds among the products of coal hydrogenations.

15.9   Pesticides

While glass capillary columns have been used in a number of fields, most analyses of chlorinated pesticides and similar compounds continue to employ packed columns. A few workers have demonstrated that the lower reactivity of the glass open tubular column, combined with its much higher powers of resolution, make it a vastly superior tool for this area of endeavor.

Franken and Rutten emphasized that even glass WCOT columns required some type of deactivation for satisfactory pesticide analysis. They obtained their best results when using apolar films of reasonable thickness, and achieved good separations of synthetic pesticide mixtues [55].

Shulte and Acker [56] also found a deactivation necessary. They deposited a heavy coating of BTPPC by passing a 1% solution in dichloromethane through the capillary in a 50°C water bath. The

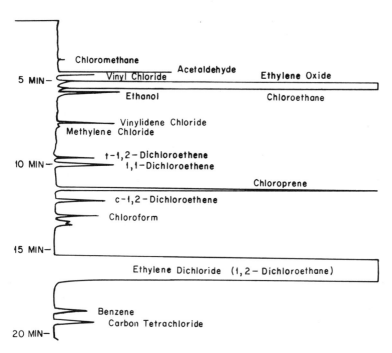

FIGURE 15.23 Low-molecular-weight chlorinated hydrocarbons. Column and conditions as in Figure 15.13. (From Johansen [53].)

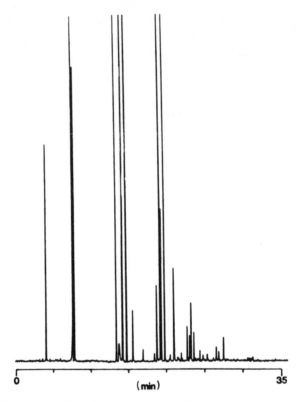

FIGURE 15.24   Capillary chromatogram of sulfur-containing compounds in coal hydrogenation products. Sulfur mode of FPD. Column, 25 m × 0.25 mm, coated with OV 101, split injection. (From Yang and Cram [54].)

excess BTPPC was then removed by washing with solvent at room temperature, leaving a thin coherent layer of the phosphonium salt. The column was then coated with SE 30, OV 101, or Dexsil 300, using a dynamic technique. They reported very good separations of a number of chlorinated hydrocarbon pesticides.

Badings *et al.* [31] used a 0.7-mm × 24-m glass WCOT column coated with a 1-μ film of SE 30, and demonstrated reasonable separation of a test series of chlorinated pesticides. The performance was much improved when they used instead a 0.9-mm × 8-m column coated with DEGS/phosphoric acid (3:1).

Holmstead [57] used glass capillary systems to follow the reaction products of the insecticide Mirex (dodecachloropenta-cyclo[5.3.0,0^{2,6}.0^{3,9}.0^{4,8}]decane) when exposed to environmental degradation.

Saleh and Casida [58] reported that each of eight different samples of toxaphene (prepared by the chlorination of camphene and related terpenes) gave the same 29 major peaks in almost identical ratios when examined by glass capillary gas chromatography (Figure 15.26). They utilized 0.25-mm × 30-m WCOT columns coated

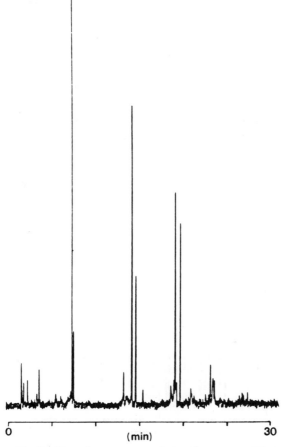

FIGURE 15.25 Capillary chromatogram of phosphorus-containing compounds in coal hydrogenation products. Phosphorus mode of FPD. Column same as Figure 15.16. (From Yang and Cram [54].)

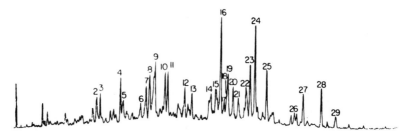

FIGURE 15.26   Glass capillary chromatogram of toxaphene. [Reprinted by permission from Saleh and Casida, *J. Agric. Food Chem.* **25**, 63 (1977). Copyright by the American Chemical Society.]

with SE 30 and found they were able to achieve a much more detailed analysis of toxaphene composition.

Improved separation of polychlorinated biphenyls was reportedly achieved by resorting to pressure programming with glass capillary columns [59]. (See also Figure 15.27.)

Buser [60] studied the applicability of glass capillary columns coated with OV 17, OV 101, and Silar 10 C to the separation of polychlorinated dibenzo-*p*-dioxins and benzofurans. He reported that thin-film coatings on narrow-bore columns showed greatly

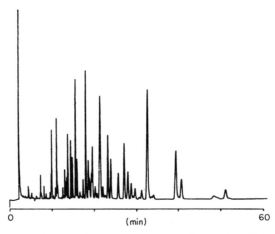

FIGURE 15.27   Glass capillary chromatogram of Arochlor 1260 on a 25-m × 0.25-mm column, coated with OV 101; split injection, isothermal at 210°C. (From Yang and Cram [54].)

increased separation efficiencies for these compounds and allowed the use of lower operating temperatures. He stressed the superiority of the glass capillary column for analyses of this type. This agrees with observations of Krupcik *et al.*, who reported that soda-lime capillaries etched with gaseous hydrogen chloride and coated with thin films of OV 101 exhibited adsorption properties that resulted in increased retention indices. Such columns were capable of separating pairs of polychlorinated biphenyls that remained unresolved when chromatographed on columns with thicker films of stationary phase [61].

Figure 15.28 shows the separation of a synthetic mixture of a number of aromatic hydrocarbons, chlorinated hydrocarbons, and pesticides [10]. A method for the determination of several organophosphorus insecticides in the serum of poisoned human patients was described by Potter *et al.* [62]. One extracts 1 cm^3 of serum with benzene, the organic layer is separated and dried, and the residue is redissolved in ethyl acetate. On-column injections utilized a 50-m × 0.5-mm column coated with SE 30. The application of glass capillary columns to pesticide analysis is the subject of a current review [63].

15.10 Saccharide Analysis

Glass capillary gas chromatography was used to determine the distribution of hydroxyethyl groups in glucose units resulting from the reaction between ethylene oxide and starch. The sample was hydrolyzed, silylated, and complete separation of the 6 monosubstituted and 12 disubstituted anomeric forms was achieved. Higher derivatives up to pentasubstituted products were separated in groups that permitted quantitative determination of the degree of substitution with a high degree of precision [64].

Szafranek *et al.* [65] reported the separation of mixtures of monosaccharide derivatives on a glass capillary coated with SE 30 admixed with a fumed silicon dioxide (Silanox 101). Schomburg and Husmann separated the silylated reaction products of gamma-irradiated glucose (through C$_6$ compounds) on WCOT glass columns [20]. The separation of the trimethylsilyl and methoxylamine–trimethylsilyl derivatives of a mixture of eight disaccharides has also been reported [66] (Figure 15.29).

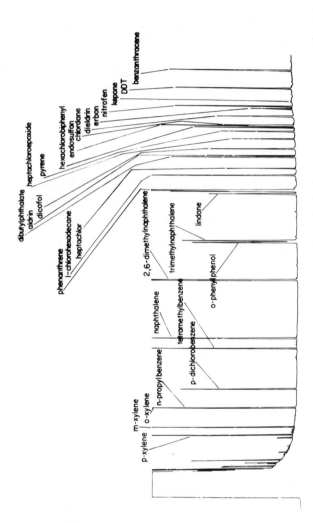

FIGURE 15.28 Chromatogram of a synthetic mixture of pollutant-type chemicals. Splitless injection, programmed from 50° to 280°C at 4°C/min on a 30-m × 0.25-mm glass capillary coated with SE 54. (From Jenkins [10].)

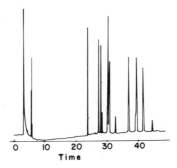

FIGURE 15.29   Glass capillary chromatogram of the MOS-TMS derivatives of a disaccharide mixture. (From Adam and Jennings [66]; reprinted by permission of the copyright owner.)

15.11   Steroid and Bile Acid Analysis

The separation of free underivatized cholesterol, campesterol, stigmasterol, and β-sitosterol was demonstrated on a 40-m × 0.25-mm column coated with Carbowax 20 M and operated isothermally at 260°C; the silylated derivatives of pregnanediol, andosterone, etiocholanolone, epiandrosterone, pregnanetriol, and cholesterin were separated on a somewhat shorter column at 270°C [20]. Sandra et al. [67] separated a variety of methoxylamine-trimethylsilyl derivatives of urinary steroids (Figure 15.30).

Novotny resolved complex mixtures of plasma steroids as their methoxime-trimethylsilyl derivatives. He employed fractionation into different conjugate groups prior to the analysis [68].

Madani et al. reported that a methylsiloxane-coated glass open tubular column gave vastly superior results in the analysis of hormonal steroids (Figure 15.31). Dichlorodimethylsilane is hydrolyzed to prepare a siloxane polymeric mixture, which is then coated on to the previously etched wall by a base-catalyzed reaction to yield a nonpolar system of high stability with good chromatographic properties. They demonstrated the separation of a wide range of sterols, steroids, and steroid metabolites [69]. The separation of a series of estrogens is shown in Figure 15.32 [70]. Figure 15.33 illustrates a glass capillary chromatogram of the TMS esters of serum-extracted bile acids [71]. Bile acid analysis [72] and steroid and hormone analysis [73] by glass capillary gas chromatography are also the topics of current reviews.

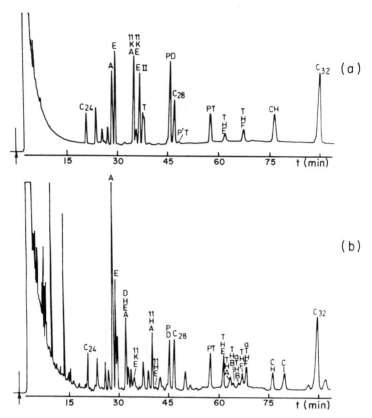

FIGURE 15.30   Glass capillary chromatograms of (a) a steroid test mixture and (b) male urinary steroids. (From Sandra *et al.* [67]; reprinted by permission of the copyright owner.)

15.12   Analyses Related to Water Quality

Recent years have seen increasing concern with the fact that levels of man-made pollutants in our water supplies are increasing at an ever-faster rate. Although a number of steps have been taken (and many others are planned) to combat this trend, any meaningful control hinges on analysis—what are the materials, what is their source, how long do they persist and under what conditions, and what are the degradation products? This is an area particularly well suited to glass capillary analysis. A compre-

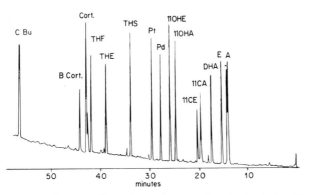

FIGURE 15.31 Glass capillary chromatogram of a standard steroid mixture. (From Madani *et al.* [69]; reprinted by permission of the copyright owner.)

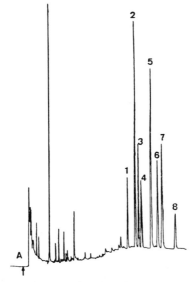

FIGURE 15.32 Results of a modified on-column injection of 50 pg of estrogen as hepta fluorobutyric ester derivatives. Column, 25 m × 0.35 mm, coated with SE 54 and programmed from 140° to 200°C at 6°C/min. Electron capture detection. (From Kern and Brander [70].)

1. 1,3,5-estratriene-3, 17-α-diol
2. 1,3,5-estratriene-3,16-β, 17-α-triol
3. 1,3,5-estratiene-3, 17-β-diol
4. 1,3,5-estratriene-3-ol, 17-one

5. 1,3,5-estratriene-3, 16-α, 17-β-triol
6. 1,3,5-estratriene-3, 16-α, 17-α-triol
7. pregnandiol
8. 1,3,5-estratriene-3, 16-β, 17-β-triol

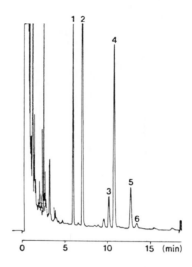

FIGURE 15.33 Chromatogram of TMS ethers of methyl esters of bile acids extracted with XAD-2 resin from 0.5-cm³ serum of a patient with liver disease. Column, 20 M coated with Carbowax 20 M, isothermal at 230°C. (From Karlaganis *et al.* [71].)

1. cholesterol
2. cholic acid
3. deoxycholic acid

4. chenodeoxycholic acid
5. hyodeoxycholic acid
6. lithocholic acid

hensive review of glass capillary chromatography as applied to water pollution analysis has recently been published [74].

Grob [12] demonstrated the applicability of glass capillary gas chromatography to the examination of sewage extracts. Giger and Schaffner [75] explored the surface layer of lake sediments in Switzerland and found an impressive array of polynuclear aromatic hydrocarbons (Figure 15.34). The authors deduced that the source of these contaminants was run-off from the city drainage. Figure 15.35 shows the direct injection of water containing two different levels of haloform additions [3]. Ether extracts of surface water are shown in Figure 15.36 [76]. Kolb *et al.* [47] used an automated headspace-injection apparatus to obtain chromatograms of several halogenated hydrocarbons in water (Figure 15.37). Figure 15.38 shows material recovered from an oil slick, resulting from a crude oil spill [10].

In the United States the deterioration of water quality resulted in the passage of a Clean Water Act, portions of which establish a National Pollutant Discharge Elimination Permit System soon scheduled to take effect. Various groups of industries will probably be required to monitor their discharges, qualitatively and quantitatively, for selected compounds. Again, the high resolving power, short analysis times, and high sensitivity of glass capillary gas chromatography offer distinct advantages, and most such testing programs will almost surely be based on this emerging technology. The separation of a standard mixture of base–neutral-extractable water pollutants is shown in Figure 15.39 [10].

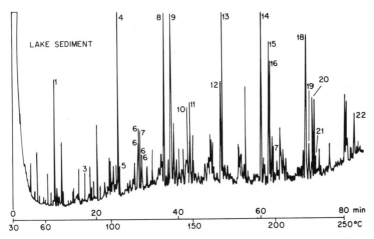

FIGURE 15.34   Gas chromatograms of polynuclear aromatic hydrocarbons isolated from the surface layer of recent lake sediments in Greifensee, Switzerland. Column, 20 m × 0.3 mm coated with SE 52, 60°–250°C at 2°C/min. (From Giger and Schaffner [75].)

1. biphenyl	12. benzo[a]anthracene
2. acenaphthene	13. chrysene/triphenylene
3. fluorene	14. benzofluoranthenes
4. phenanthrene	15. benzo[e]pyrene
5. anthracene	16. benzo[a]pyrene
6. methylphenanthrenes	17. perylene
7. 4,5-methylene phenanthrene	18. dibenzanthracenes
8. fluoranthene	19. indeno[1,2,3-cd]pyrene
9. pyrene	20. benzo[g,h,i]perylene
10. benzy[a]fluorene	21. anthracene
11. benzo[b]fluorene	22. coronene

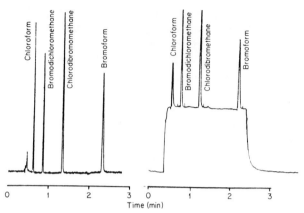

FIGURE 15.35   Left, split injection of water containing ~15 ng/μl of the indicated halomethanes, FID. Right, ~15 pg/μl, ECD. Column, 10 m × 0.25 mm, coated with SP 2100; 50°C isothermal. (From Rooney *et al.* [3].)

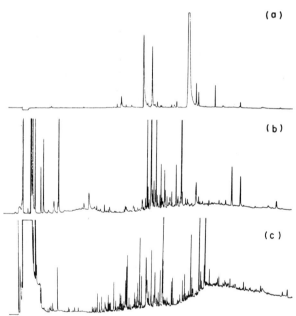

FIGURE 15.36   Ether extract of surface water analyzed via three detection modes: (a) N-P FID, (b) ECD, and (c) FID. Column, 50 m × 0.25 mm, coated with OV 101, programmed from 25° to 250°C at 4°C/min. (From Rooney and Freeman [76].)

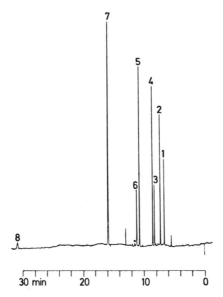

FIGURE 15.37   Headspace analysis of halocarbons in water. Column, 43 m ×
0.2 mm coated with UCON LB 550, 85°C isothermal. (After Kolb *et al.* [47].)

1. dichloromethane
2. carbon tetrachloride
3. 1,2-dichloroethane
4. trichloroethylene

5. tetrachloroethylene
6. bromodichloroethane
7. chlorobenzene
8. bromoform

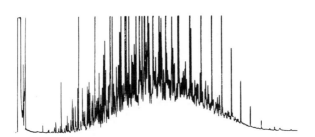

FIGURE 15.38   Material from an oil slick at sea. Column, 30 m × 0.25 mm,
coated with SP 2100 and programmed from 100° to 280°C at 4°C/min. Pattern is
typical of a crude oil, although crudes can sometimes be differentiated on the
basis of chromatographic patterns. (From Jenkins [10].)

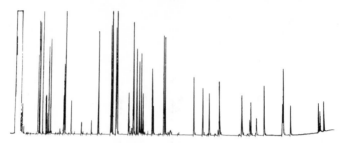

FIGURE 15.39   Base neutral-extractable water pollutants, standard mixture. Column, 15 m × 0.24 mm SE 54, programmed 30°–280°C at 4°C/min. Hydrogen carrier at $\bar{u}_{\text{initial}}$ 62.5 cm/sec and $\bar{u}_{\text{final}}$ 40 cm/sec. Splitless injection, 2 μl sample. (From Jenkins [10].) Components: methanol and 2-chloroethyl vinyl ether; N-nitrosodimethylamine bis(2-chloroethyl) ether; 1,3-dichlorobenzene; 1,4-dichlorobenzene; 1,2-dichlorobenzene; bis(2-chloroisopropyl) ether; hexachloroethane; N-nitroso-di-n-propylamine; nitrobenzene; isophorone; bis(2-chloroethyoxy) methane; 1,2,4-trichlorobenzene; naphthalene; hexachlorobutadiene; hexachlorocyclopentadiene; 2-chloronaphthalene; acenaphthylene; dimethyl phthalate; 2,6-dinitrotoluene; acenaphthene; 2,4-dinitrotoluene; fluorene; diethyl phthalate; N-nitrosodiphenylamine; 1,2-diphenylthydrazine (as azobenzene); 4-bromophenyl-phenyl ether; hexachlorobenzene; phenanthrene; anthracene; di-n-butyl phthalate; fluoranthene; pyrene; benzidine; endosulfan sulfate; butylbenzyl phthalate; benz(a)anthracene; chrysene; 3,3′-dichlorobenzidine; bis (2-ethylhexyl)phthalate; din-octyl phthalate; benzo(b)fluoranthene; benzo(k)fluoranthene; benzo(a)pyrene; indeno(1,2,3-cd)pyrene; dibenzo (a,h)anthracene; benzo(ghi)perylene.

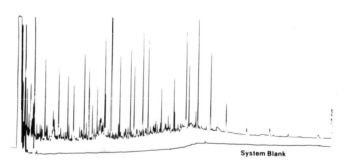

FIGURE 15.40   Results from an automated analysis of Delaware River water, base–neutral extract. Column, 35 m × 0.25 mm, coated with SP 2250; 2 min 80°C, programmed at 3.5°C/min to 220°C; helium carrier gas at 30 cm/sec. Splitless injection. (From Rooney and Freeman [77].)

TABLE 15.1

Computer Scanning for Priority Pollutants, Delaware River—Base Neutral Extract[a]

EPA limit test for sample No. 5. SP-2100 column indicates possibility of:

Name	RT	Amount (NG)
Acenaphthene	17.477	33.10
Acenaphthyle	16.404	12.08
Anthracene	27.391	2.84
Dibutyl PH	33.083	26.01
Diethyl PH	21.921	3.46
Fluoranthene	35.103	1.32
Fluorene	20.869	0.09
Naphthalene	8.012	15.13
Phenanthrene	27.079	1.21

Run in progress on other channel

Results from SP-2250 Column:

Name	RT match	Amount (NG)	Confirmed	Exceeds limit
Acenaphthene	No		No	
Acenaphthyly	Yes	1.32	Yes	No
Anthracene	Yes	1.21	Yes	No
Dibutyl PH	Yes	25.97	Yes	Yes
Diethyl PH	No		No	
Fluoranthene	No		No	
Fluorene	Yes	2.80	Yes	No
Naphthalene	Yes	15.40	Yes	No
Phenanthrene	No		No	

[a] The retention time windows for selected pollutant standards are established on two dissimilar high-resolution columns (results from one analysis are shown in Figure 15.37). On the basis of these data, the final report prints out an analysis for those selected pollutants [77].

One approach that has been suggested for the control of these priority pollutants in industrial wastes is based on their retention characteristics on two high-resolution glass capillary columns containing dissimilar liquid phases (Section 7.6). Computer analysis of the results of sample injection on one column indicate which of the "windows" (assigned by precalibration with standards) are occupied by peaks; computer analysis of the results from the second column either rule out that assignment or confirm it [77]. Figure 15.40 illustrates the technique as applied to a sample of river water, and Table 15.1 shows a computer analysis of those data.

Because of the very thin wall, fused silica columns exhibit much smaller outer diameters, and the inlet ends of two or more columns can easily be bunched together for connection into a common inlet, while the outlet of each column can be connected to a separate detector.

Phenols, which are widely regarded as critically important priority pollutants, can be especially difficult to analyze. Because they exhibit even higher affinities for active sites, halogen and nitro-substituted phenols are even more difficult, and total or partial abstraction, exhibited by malformed and badly tailing peaks, has been the rule rather than the exception. Fused silica columns offer great promise in this direction, as illustrated by Figure 15.41.

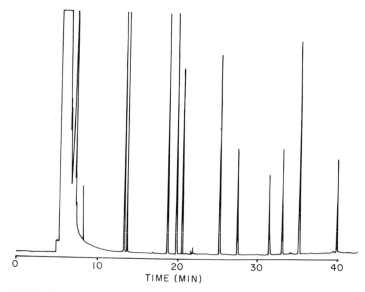

FIGURE 15.41 Priority pollutant analysis of underivatized phenols on a 0.22 mm = 15 m fused silica column coated with SE 30 and programmed 100°–220° at 8°C/min. Compounds include 2-chlorophenol, phenol, dimethylphenol, 2-nitrophenol, dichlorophenol, dichlorocresol, trichlorophenol, p-nitrophenol, dinitrophenol, dinitromethylphenol, and pentachlorophenol. (Courtesy of J&W Scientific.)

References

1. Novotny, M., McConnell, M. L., and Lee, M. L., *J. Agric. Food Chem.* **22**, 765 (1974).
2. Johansen, N., *Chromatogr. Newsl.* **5**(1), 14 (1977).
3. Rooney, T. A., Altmayer, L. H., Freeman, R. R., and Zerenner, E. H., *Am. Lab.* **11**(2), 81 (1979).
4. Lee, M. L., and Hites, R. A., *Anal. Chem.* **48**, 1890 (1976).
5. Lee, M. L., Novotny, M., and Bartle, K. D., *Anal. Chem.* **48**, 405 (1976).
6. Novotny, M., McConnell, M. L., and Lee, M. L., *J. Agric. Food Chem.* **22**, 765 (1974).
7. Lee, M. L., Novotny, M., and Bartle, K. D., *Anal. Chem.* **48**, 1566 (1976).
8. Rapp, U., Schröder, U., Meier, S., and Elmenhorst, M., *Chromatographia* **8**, 474 (1975).
9. Onuska, F. I., and Comba, M. E., *J. Chromatogr.* **126**, 133 (1976).
10. Jenkins, R., personal communication (1979).
11. Roeraade, J., *Chromatographia* **8**, 511 (1975).
12. Grob, K., *Chromatographia* **8**, 423 (1975).
13. Hoshika, Y., and Takata, Y., *J. Chromatogr.* **120**, 379 (1976).
14. Novotny, M., Lee, M. L., Low, C., and Raymond, A., *Anal. Chem.* **48**, 24 (1976).
15. Bjørseth, A., and Eklund, G., *HRC&CC* **2**, 22 (1979).
16. Bertsch, W., *in* "Glass Capillary Gas Chromatography. The Applications" (W. Jennings, ed.). Dekker, New York, 1980. In press.
17. Deyl, Z., *J. Chromatogr.* **127**, 91 (1976).
18. Eyem, J., and Sjøquist, S., *Anal. Biochem.* **52**, 255 (1973).
19. Cavadore, J. C., Nota, G., Prota, G., and Previero, A., *Anal. Biochem.* **60**, 608 (1974).
20. Schomburg, G., and Husmann, H., *Chromatographia* **8**, 517 (1975).
21. Adams, R. F., Vandemark, F. L., and Schmidt, G. J., *J. Chromatogr. Sci.* **15**, 63 (1977).
22. Poole, C. F., and Verzele, M., *J. Chromatogr.* **150**, 439 (1978).
23. Abe, I., Kohno, T., and Musha, S., *Chromatographia* **11**, 393 (1978).
24. Nicholson, G. J., Frank, H., and Bayer, E., *HRC&CC* **2**, 411 (1979).
25. Grob, K., and Grob, K., Jr., *J. Chromatogr.* **151**, 311 (1978).
26. Grob, K., *HRC&CC* **1**, 263 (1978).
27. Jaeger, H., Frank, H., Klör, H. U., and Ditschuneit, H., *in* "Glass Capillary Gas Chromatography. The Applications" (W. Jennings, ed.). Dekker, New York, 1980. In press.
28. Rijks, J. A., and Cramers, C. A., *Chromatographia* **8**, 482 (1975).
29. Dünges, W., Langlais, R., and Schlenkermann, R., *HRC&CC* **2**, 361 (1979).
30. Sirikulvadnana, S., Jennings, W. G., and Vogel, G., *Int. Flavours* Mar./Apr., p. 126 (1975).
31. Badings, H. T., van der Pol, J. J. G., and Wassink, J. G., *Chromatographia* **8**, 440 (1975).
32. Jaeger, H., Klör, H., Blos, G., and Ditschuneit, H., *J. Lipid Res.* **17**, 185 (1976).

33. van Vleet, E. S., and Quinn, J. G., *J. Chromatogr.* **151**, 396 (1978).
34. Grob, K., Jr., Neukom, H. P., Frölich, D., and Battaglia, R., *HRC&CC* **1**, 94 (1978).
35. Jaeger, H., Klör, H. U., and Ditschuneit, H., in "Glass Capillary Gas Chromatography. The Applications" (W. Jennings, ed.). Dekker, New York, 1980. In press.
36. Shibamoto, T., and Russell, G. F., *J. Agric. Food Chem.* **24**, 843 (1976).
37. Uchman, W., and Jennings, W. G., *J. Food Chem.* **2**, 135 (1977).
38. Jennings, W. G., and Filsoof, M., *J. Agric. Food Chem.* **25**, 440 (1977).
39. Jennings, W. G., *J. Food Chem.* **2**, 185 (1977).
40. Yabumoto, K., Jennings, W. G., and Yamaguchi, M., *J. Food Sci.* **42**, 32 (1977).
41. Yamaguchi, M., Hughes, D. L., Yabumoto, K., and Jennings, W. G., *Sci. Hortic.* **6**, 59 (1977).
42. Yabumoto, K., Yamaguchi, M., and Jennings, W. G., *Chem., Mikrobiol., Techol. Lebensm.* **5**, 53 (1977).
43. Jennings, W. G., and Wohleb, R., *Chem., Mikrobiol., Technol. Lebensm.* **3**, 33 (1974).
44. Kugler, E., and Langlais, R., *Chromatographia* **8**, 468 (1975).
45. Jennings, W. G., and Bernhard, R. A., *Chem., Mikrobiol., Technol. Lebensm.* **4**, 95 (1975).
46. Klimes, I., and Lamparsky, D., in "Analysis of Food and Beverages. Headspace Techniques" (G. Charalambous, ed.), p. 95. Academic Press, New York, 1978.
47. Kolb, B., Pospisil, B., Borath, T., and Auer, M., *HRC&CC* **2**, 261 (1979).
48. Bjørseth, A., in "Carcinogenesis. Vol. III: Polynuclear Aromatic Hydrocarbons" (P. W. Jones and R. I. Freudenthal, eds.), p. 75. Raven, New York, 1978.
49. Meili, J., Brönnimann, P., Brechbühler, B., and Heiz, H. J., *HRC&CC* **2**, 475 (1979).
50. Mathews, R. G., Torres, J., and Schwartz, R. D., *HRC&CC* **1**, 139 (1978).
51. Averill, W., and March, E. W., *Chromatogr. Newsl.* **4**(2), 20 (1976).
52. Dielmann, G., Meier, S., and Rapp, A., *HRC&CC* **2**, 343 (1979).
53. Johansen, N. G., *Chromatogr. Newsl.* (to be published).
54. Yang, F. J., and Cram, S. P., *HRC&CC* **2**, 487 (1979).
55. Franken, J. J., and Rutten, G. A. F. M., in "Gas Chromatography 1972" (S. G. Perry, ed.), p. 75. Appl. Sci., New York, 1973.
56. Shulte, E., and Acker, L., *Z. Anal. Chem.* **286**, 260 (1974).
57. Holmstead, R., *J. Agric. Food Chem.* **24**, 620 (1976).
58. Saleh, M. A., and Casida, J. E., *J. Agric. Food Chem.* **25**, 63 (1977).
59. Mourits, J. W., Merkus, H. G., and de Galan, L., *Anal. Chem.* **48**, 1557 (1976).
60. Buser, H. R., *Anal. Chem.* **48**, 1553 (1976).
61. Krupcík, J. Kristín, M., Valachovicová, M., and Janiga, S., *J. Chromatogr.* **126**, 147 (1976).
62. Potter, M. E., Muller, R., and Willems, J., *Chromatographia* **11**, 220 (1978).
63. Hermann, B. W., and Seiber, J. N., in "Glass Capillary Gas Chromatography. The Applications" (W. Jennings, ed.). Dekker, New York, 1980. In press.
64. Mourits, J. W., Merkus, H. G., and de Galan, L., *Anal. Chem.* **48**, 1557 (1976).

65. Szafranek, J., Pfaffenberger, C. D., and Horning, E. C., *J. Chromatogr.* **88**, 149 (1974).
66. Adam, S., and Jennings, W. G., *J. Chromatogr.* **115**, 218 (1975).
67. Sandra, P., Verzele, M., and VanLuchene, E., *Chromatographia* **8**, 499 (1975).
68. Novotny, M., Maskarinec, M. P., Steverink, A. T. G., and Farlow, R., *Anal. Chem.* **48**, 469 (1976).
69. Madani, C., Chambaz, E. M., Rigaud, M., Durand, J., and Chebroux, P., *J. Chromatogr.* **126**, 161 (1976).
70. Kern, H., and Brander, B., *HRC&CC* **2**, 312 (1979).
71. Karlaganis, G., Paumgartner, G., and Schwarzenbach, R. P., *HRC&CC* **2**, 293 (1979).
72. Jaeger, H., Nebelung, W., Klör, H. U., and Ditschuneit, H., *in* "Glass Capillary Gas Chromatography. The Applications" (W. Jennings, ed.). Dekker, New York, 1980. In press.
73. Vanluchene, E., and Sandra, P., *in* "Glass Capillary Gas Chromatography. The Applications" (W. Jennings, ed.). Dekker, New York, 1980. In press.
74. Lin, D. C. K., *in* "Glass Capillary Gas Chromatography. The Applications" (W. Jennings, ed.). Dekker, New York, 1980. In press.
75. Giger, W., and Schaffner, C., *Anal. Chem.* **50**, 243 (1978).
76. Rooney, T. A., and Freeman, R. R., Tech. Pap. No. 69, Hewlett Packard Co. (1979).
77. Rooney, T. A., and Freeman, R. R., Tech. Pap. No. 83, Hewlett Packard Co. (1979).

CHAPTER 16

FAULT DIAGNOSIS

16.1 General Considerations

In the vast majority of cases the operator is in a good position to diagnose problems and take any necessary remedial action, but it is important to recognize at what point professional assistance should be requested. Most users are simply not qualified to correct electrical or electronic faults, and by the same token, few servicemen are currently well versed in glass capillary GC. However, it can be of considerable assistance to the serviceman if, when he requests assistance, the user is able to describe a logical diagnostic approach that he employed to conclude, for example, that the electrometer itself was defective.

Coverage in this chapter is limited to problems with which the average chromatographer should be able to cope. Our diagnoses will be based on the recorder trace; to ensure that the problem is associated with generation or processing rather than recording of the signal, it is sometimes informative to wire a second recorder in parallel with the first. Noise spikes can be due to a dirty slide wire, and malformed peaks can result from a recorder pen that moves sluggishly (sometimes restricted to either the up or down direction), a defect that can be of electrical or mechanical origin. The manufacturer's literature usually gives detailed instructions for the routine maintenance of the recorder, including lubrication and slide wire cleaning. To check for mechanical sluggishness on a standard strip chart recorder, disconnect the recorder power and ensure that the pen can be moved freely by hand from one extreme to the other, with no evidence of binding. *Caution:* Some thermal printers can be damaged if the print head is moved manually; the manufacturer's literature should be consulted. Many recorders require periodic lubrication of the drive cable pulleys, servo motors, and pen sliders; others utilize bearings and sliders that are not designed for lubrication and that build up gummy deposits and become sluggish with the application of mineral oils. Again, the manufacturer's literature should be consulted.

16.2   The Methane Injection

If the methane injection fails to produce a response, first verify that the recorder is connected, that the electrometer sensitivity settings (input and attenuation) are appropriate, and that the electrometer is functional. All of these can usually be checked by rotating the zero or baseline adjust control on the electrometer and noting that pen response is normal. With FID, check that the flame is lighted; a convenient method is to hold a cold surface (mirror, beaker, spatula, etc.) at the FID outlet and check for condensation. In most cases this will also produce a signal perturbation that verifies that the FID in question is in fact connected through the electrometer to the recorder; with instruments designed to handle a multiplicity of inputs and/or outputs, this is sometimes a problem.

The lack of response to an injection can also result from a clogged or defective syringe. Ensure that the needle is not plugged and that the syringe displaces air by blowing bubbles into a beaker of water. Clogs in the standard 10 μl syringe most commonly occur within the needle, and are best corrected by passing a suitable cleanout wire through the needle. A clean syringe should then be used to insert 5–10 μl of a suitable solvent into the barrel of the original syringe. With the plunger removed, the standard 28 gauge needle of the second syringe can be inserted into the glass bore of the syringe being cleaned. The plunger is then fitted and solvent forced through the needle. Sufficient pressure can be developed to split the glass barrel if a cleanout wire has not been first passed through the needle.

Prompt and proper cleaning of syringes can help avoid these problems. Syringes are best cleaned with solvent flowing from the barrel through the needle rather than in the reverse direction. Debris such as small pieces of septa are removed more easily by flushing in this direction, while movement in the opposite direction can cause them to lodge more deeply within the needle. Laboratories equipped with plumbed vacuum sources find it convenient to place a short length of plastic tubing with a serum cap over one end on the vacuum jet. The syringe is cleaned by removing the plunger, inserting the needle through the serum cap, and using a clean syringe or a micropipet to introduce several bursts of a suitable solvent through the syringe barrel. Initial solvents should be selected on the basis of their ability to solubilize any residues from the samples used; acetone, acetonitrile, pyridine, methanol, or water may be more suitable. The final solvent should be a low boiler such as acetone or dichloromethane, after which air is allowed to pass through the syringe for a few moments. The plunger is best cleaned by wiping it down with a lintless tissue moistened with a suitable solvent, and finally with a tissue moistened with dichloromethane. Tungsten plunger syringes, in which the sample is contained within the needle, and syringes fitted with special fine (e.g., 32 gauge) needles cannot be cleaned by the above procedure, and the manufacturer's recommendations should be consulted. Another problem is syringe blow-by. To check for this defect, substitute $0.3-0.5$ μl of a low-boiling compound (isopentane, pentane, dichloromethane) and with that material within the glass syringe barrel, insert the needle into the inlet while carefully observing the sample in the syringe. A worn or mismatched plunger and syringe may permit carrier gas to blow by, forcing the sample past the plunger and filling the syringe with carrier gas; the injection is then merely carrier gas. Also check that the split ratio is not excessive (Section 4.4) and that the septum is not leaking. Check make-up, hydrogen, air, and carrier gases; an exhausted air or make-up tank will lead to a decrease in the rate of delivery of these to the FID and can have an adverse effect on sensitivity. Check for leaks at the column connections or whether the column has broken; verify that there is flow through the column (Section 5.3).

The methane (or propane, see Section 14.2) peak should be needle sharp. Peak width will, of course, be influenced by the recorder chart speed; even a poor peak can be made to look good by resorting to a low chart speed. As a rough rule of thumb, with a 50-m column, proper gas velocity, and acceptable injection technique, a methane peak achieving full scale at normal operational sensitivity that exceeded 1 sec in baseline width should be regarded as unsatisfactory. At a chart speed of 2 cm/min, this amounts to 0.3 mm.

Three examples are shown in Figure 16.1. The first example illustrates a peak of unacceptable width, which also possesses a tail. This indicates that the injection occurred over too long a period of time and probably relates to faults in the inlet, too low

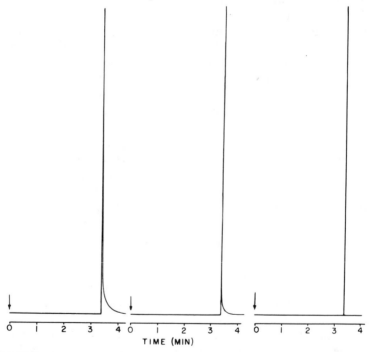

TIME (MIN)

FIGURE 16.1 The methane injection; typical results on a 50-m capillary at $\bar{u}$ $\approx$ 20 cm/sec; chart speed 2 cm/min. Left, unacceptable peak width, plus tailing. Center, good peak width, but tailing. Right, ideal methane peak. See text for details.

a split ratio, or insufficient make-up gas. The second example has an acceptable width but must be faulted for tailing. A butt joint (or a break) resulting in a minute dead volume at either end of the column could cause this. The third example illustrates a well-formed methane peak.

16.3   Test Mixture Injection

All the examples shown in Figure 16.2 assume the split injection of a test mixture, although conclusions can in many cases be carried forward to other injection modes. Example A shows a normal chromatogram of that test mixture, to which the other examples should be compared. Example B shows a situation where later peaks exhibit an asymmetry that is characterized by a "leading tail"; less obvious cases of this are easier to diagnose by observing the fact that the recorder pen rises slowly and returns to baseline more rapidly. This is the result of a sample overload, which can be corrected by decreasing the sample size (and raising sensitivity) or increasing the column temperature; the latter route corrects the overload by forcing more of the solute into the gas phase. Peaks that exhibit overloading (or any other asymmetry) are not chromatographing under equilibrium conditions and should not be used, e.g., for efficiency measurement calculations.

Example C shows good peak conformation with most solutes, but severe tailing of an active compound, perhaps an alcohol. This indicates that adsorptive sites have been generated or exposed. While these may occur in the column, they may also be in the inlet or detector. Many good columns have been condemned for faults that existed in the inlet or detector. Normally, such defects can be rectified (Sections 5.3, 5.5). In example D all peaks are atypically broad, reflecting a loss of system efficiency. This may be due to a variety of causes. The carrier gas velocity, the split ratio, and the flow rate of make-up gas should all be checked. This symptom can also be caused by a cracked inlet liner or a column break at the attachment points (sometimes within the ferrule). A tiny amount of graphite, Vespel, or other material, scraped from the ferrule as it is slipped over the column, or a single glass shard within the column will lead to tailing and/or severe losses in system efficiency. Crumbs of silicone rubber may

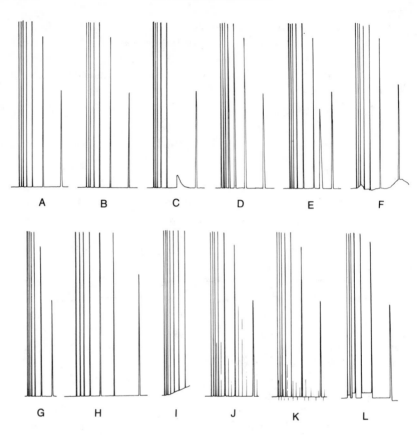

FIGURE 16.2   Test mixture injections. (A) ideal, (B) overloaded (leading tail), (C) adsorption and/or degradation of an active compound, (D) loss of system efficiency, (E) extraneous peak, (F) meandering baseline, (G) atypically short retentions, (H) atypically long retentions, (I) steep baseline rise on programming, (J) positive noise spikes, (K) random noise spikes, (L) stepped baseline. See text for details.

have accumulated in the inlet, or the column may be deteriorating (Chapter 10). In some cases efficiency can be restored by removing or cleaning a short section of the column (Section 10.6). Short sections of column broken off in the inlet or detector can also lead to serious tailing problems. Where the broken segment lodges in such a position that a portion of the flow stream is diverted through it, doublet peaks can result (Sections 4.8 and 14.7).

An extra peak has appeared in example E. Such peaks can result from a number of sources (e.g., Section 4.2), but in this case the width of that peak at half height exceeds the normal progression of this value with increased partition ratio (Section 6.3), indicating that it did not originate from this same injection. Example F exhibits a normal chromatogram, but the baseline meanders and considerable drift is evident. In the majority of cases these defects are caused by higher-boiling compounds that may have spent days on column and are finally being eluted. A stable baseline may be recovered if the column is cooked out by raising the temperature for a period of time. The use of higher temperatures will accomplish this more rapidly but may also have an adverse effect on the column deactivation treatment and column life (Section 10.3). Drifting and/or noisy baselines can also relate to the need to replace molecular sieve filters (and probably to clean or replace associated gas lines, valves, etc.; contaiminated lines and fittings downstream from the filters will continue to give problems). In addition, baseline disturbances of this type can relate to higher-boiling materials from the septum or to injection residues in the inlet or the column; the latter can sometimes be rectified by removing or cleaning that section of the column (Section 10.6).

Example G shows a decrease in partition ratios of all solutes; that this is not merely a case of a higher carrier gas velocity is attested to by the fact that t_M (estimated in this case from the leading edge of the solvent peak) has remained constant. This probably relates to a loss of liquid phase, resulting in a higher phase ratio. In example H all retentions are longer; if t_M had remained constant, such a chromatogram would probably indicate a polarity shift (Section 10.5), but in this case t_M has also increased and partition ratios agree with those on the earlier chromatograms. Hence this change must relate to a shift in carrier gas velocity; it is probable that the pressure in the inlet has decreased owing to a lower supply pressure, a massive septum leak, or a leak at the column inlet connection.

Example I exhibits a relatively steep baseline rise (temperature programming); with a silicone phase, one would question the type of glass used in column construction (Section 2.2); this may also be caused by deactivation chemicals leaving the column. In some cases where these higher temperatures are required, it may be possible to select a more stable liquid phase (Section 11.1); in other cases a higher carrier gas velocity may permit the use of lower temperatures (Sections 8.5, 8.6).

Examples J and K illustrate spiking and noise problems, but in example J all of the extraneous signal is positive-going, whereas in K some is also negative-going. In example J the detector receives additional spurts of signal; this may be due to a dirty detector. Condensed materials on critical detector components may be causing electrical leakage, or higher-boiling substances may have condensed within the detector plumbing, and fragments from this condensate are occasionally swept up by the high-velocity gas stream. Contaminated air, hydrogen, or make-up gas or a host of other causes can produce this result; somehow contaminants are randomly entering the detector. In one case called to the author's attention the noise was violent and close spaced and was eventually traced to solvent from a nearby HPLC that was discharged into a sink whence the outlet for forced-air ventilation sucked vapor-contaminated air past the FID.

In example K the problem is probably electronic. It is advisable to disconnect first the electrometer cable at the FID and to determine the effect, if any. If the baseline is now steady, the problem is in the detector and may be entering by way of the column or the gas supplies; check by substitution. If the baseline remains noisy, disconnect the cable where it enters the electrometer; noisy cables are not unknown. If the problem is associated with the cable itself, first try rerouting the cable; sometimes the connectors can be cleaned with carbon tetrachloride and a small brush. The abrasive rubber of a pencil-type typewriter eraser is an excellent contact cleaner; it should be used dry and infrequently (gold-plated contacts can be damaged by excessive abrasion). In many cases, however, such cables must be replaced. Check whether the noise is associated with the oven fan; an out-of-balance system may cause vibrations that lead to electronic noise. (Such vibrations may also be contributed by other apparatus on the same bench.) Noisy switch contacts can also cause this problem; gently tap the knobs, and where the problem seems to be associated with a switch or control, work that switch (or control) several times, preferably after applying a suitable contact cleaner. Spray cans of the latter, which should be used sparingly, are available at most electronic stores.

Example L illustrates a stepping problem, where the recorder abruptly varies its baseline, usually on returning from a peak. Although this may relate to a sluggish pen drive (Section 16.1), it is sometimes caused by a grounding problem, and experimentation with a small ceramic capacitor may help. The position of the capacitor should be varied, from one recorder input terminal to ground, the other terminal to ground, and across the two terminals.

16.4 Solvent Peak in Split Injection

The solvent peak quite often exhibits a tail even in the split injection mode, a phenomenon that may be accentuated by larger injections or low split ratios. This probably relates to the fact that the solvent is usually low boiling and comprises 50–90% of the injection. As this large amount of material undergoes flash vaporization, it saturates the whole injector volume. Solvent may condense in cooler areas such as the septum face; in extreme cases the solvent flash may extend to unheated portions, such as the carrier gas line. In either case that condensed solvent slowly evap-

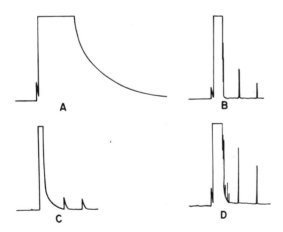

FIGURE 16.3 Fault diagnosis in splitless injection. (A) Purge function not activated, activated too late, or purge ineffective. (B) Purge function activated too soon. (C) Column temperature too high with respect to boiling point of solvent; no solvent effect. (D) Well-executed splitless injection. See text for details.

orates into the carrier gas stream, where its concentration diminishes slowly, producing a solvent tail. A septum bleed cap (Section 4.2) may help control this problem.

16.5 Splitless Injection

Figure 16.3 illustrates chromatographic traces common to the splitless mode of injection (Section 4.3). In example A the purge function was either inoperative or activated too late; the solvent tail has obliterated most of the region of interest. In example B the purge function was activated too soon; the solvent effect occurred (observe the reverse solvent effect on the leading peak, and the normal solvent effect on the solvent "impurities"), but a relatively small amount of sample moved from the inlet to the column; most of the sample was purged to atmosphere. The solvent effect did not occur in example C; this probably resulted from too high a column temperature relative to the boiling point of the solvent. Example D shows a normal well-executed splitless injection.

NOMENCLATURE

Considerable confusion can arise over gas chromatographic terminology, in that different authors have used different symbols for the same term, and even more confusing, the same symbol for different terms. For example, Littlewood [1] uses β for the distribution constant, while Ettre [2] uses the same symbol for the phase ratio of the column.

Insofar as possible, the terminology used in this offering complies with or is analogous to the recommendations of the International Union of Pure and Applied Chemistry [3], with a few minor modifications and some necessary additions.

Some special mention of retention measurements is probably in order. Retention measurements, as well as measurements of the hold-up volume and peak width, may be made in terms of times, chart distances, or gas volumes. Although volumes are preferred for most theoretical discussions, times or chart distances are easier to use in most practical situations. If chart speed and gas flow are constant, gas volume is directly proportional to time or chart distance. Because many recorders exhibit considerable backlash in the chart drive gear train, chart distance can be an unreliable measurement in the hands of the unwary. Particularly if the chart paper has been advanced manually, a finite period is required for the chart drive motor to take up the backlash and begin the chart advance.

$\alpha_{A,B}$, Relative retention

This is specifically defined as the ratio of the distribution coefficients of two substances A and B measured under identical conditions. It can be shown that $\alpha_{A,B} = t'_{R(B)}/t'_{R(A)}$. By convention, α is always greater than unity.

β, Phase ratio

The volume of the column occupied by the gas phase relative to the volume occupied by the stationary phase:

$$\beta = \frac{V_G}{V_L}$$

f_t, Film thickness

The thickness, usually in microns (μ), of the film of stationary phase.

h, Height equivalent to a theoretical plate (HEPT)

The column length divided by the theoretical plate number, usually expressed in millimeters; when measured at the van Deemter minimum u_{opt}, it is sometimes given the symbol h_{min}.

H, Height equivalent to an effective theoretical plate (HEEPT)

The column length divided by the effective theoretical plate number, usually expressed in millimeters; when measured at the van Deemter minimum u_{opt}, this may be given the symbol H_{min}.

I, Retention index

A number, obtained by logarithmic interpolation, relating the adjusted retention volume of a component A to the adjusted retention volumes of the n-paraffin hydrocarbons. Each n-paraffin hydrocarbon is arbitrarily allotted by definition an index 100 times its carbon number.

$$I_A = 100N + 100n\,\frac{\log V'_{R(A)} - \log V'_{R(N)}}{\log V'_{R(N+n)} - \log V'_{R(N)}}$$

where $V'_{R(N+n)}$ and $V'_{P(N)}$ are the adjusted retention volumes of n-paraffin hydrocarbons of carbon number N and $(N + n)$ that are, respectively, smaller and larger than $V'_{R(A)}$ [3].

In practice, most workers utilize retention times, and include reference to the liquid phase and temperature as an integral part of I:

$$I_b^a = 100N + 100n \frac{\log t'_{R(A)} - \log t'_{R(N)}}{\log t'_{R(N+n)} - \log t'_{R(N)}}$$

where I is the retention index on liquid phase a at temperature b, and the adjusted retention times are analogous to the adjusted retention volumes above.

k, Partition ratio

A measure of how long a solute spends in the stationary phase relative to the time it spends in the gas phase: $k = t_R'/t_M$. Sometimes it is termed the "capacity ratio."

K_D, Distribution constant

The ratio of the concentration of a component in a single definite form in the stationary phase to its concentration in the same form in the mobile phase at equilibrium. Both concentrations are calculated per unit volume of the phase. This term is recommended in preference to the term "partition coefficient," which has been used with the same meaning [3].

L, Length of the column

The total length of the coated portion of the column, which in a glass capillary usually extends from its inlet end to the detector end, but excluding any connecting tubing.

n, Theoretical plate number

A number used to indicate column performance calculated from the relationship

$$n = 16\left(\frac{t_R}{w}\right)^2 = 5.54\left(\frac{t_R}{w_{0.5}}\right)^2 \qquad \text{(Section 1.3)}$$

N, *Effective theoretical plate number*

A number used to indicate column performance that is based on adjusted retention measurements and that takes resolution into account the gas hold-up volume of the system:

$$N = 16\left(\frac{t_R'}{w}\right)^2 = 5.54\left(\frac{t_R'}{w_{0.5}}\right)^2$$

n_{real}, *Real theoretical plate number*

A number used to indicate total system performance, based on the calculated $w_{0.5}$ values for hypothetical substances with partition coefficients $k = 0$ and $k = 10$ (i.e., b_0 and b_{10}), as determined from the best-fit-line for a series of experimental points where $w_{0.5}$ is plotted as a function of k (Section 6.3):

$$n_{real} = 5.54\left(\frac{t_{R(10)}'}{b_{10} - b_0}\right)^2$$

R_s, *Peak resolution*

The separation of two peaks in terms of their average peak width (Figure 1.11):

$$R_s = \frac{2(t_{R(B)} - t_{R(A)})}{w_{(A)} + w_{(B)}}$$

Since $w = 1.699w_{0.5}$,

$$R_s = 1.18\frac{t_{R(B)} - t_{R(A)}}{w_{0.5(A)} + w_{0.5(B)}}$$

r_0, *Radius of the capillary*

The inner radius, usually determined with an occular micrometer or microscope.

t_M, *Hold-up time*

The time (or chart distance) required to conduct a nonsorbed component through the column; equivalent to an air-peak time, this is usually estimated by methane injection (Section 1.2) or calculated (Section 7.2).

t_R, *Retention time*

The time (or chart distance) from the point of effective injection to the peak maximum (see Figure 1.2). The peak should exhibit good symmetry; poor injection techniques, insufficient inlet or column temperature, adsorptive tailing, or overloading can adversely affect the symmetry and the position of the peak maximum.

t_R', *Adjusted retention time*

The time (or chart distance) from the point of effective injection to the peak maximum, minus the hold-up time: $t_R' = t_R - t_M$, therefore equivalent to the time a component spends in the stationary phase.

TZ, *Separation number*

A measure of column efficiency, based on the degree to which two members of an homologous series are separated, which takes into account the sharpness of the peaks and their relative retentions:

$$TZ = \frac{t_{R(B)} - t_{R(A)}}{w_{0.5(A)} + w_{0.5(B)}} - 1$$

$\bar{u}$, *Average linear velocity of the carrier gas*

Usually determined by dividing the column length in centimeters by the time in seconds required for a methane peak, t_M.

u_{opt}, *Optimum gas velocity*

That velocity reflected by the minimum in the van Deemter curve (Figure 1.13).

w, *Width of the peak at baseline*

The distance, usually in millimeters of chart or in time, from the beginning of the peak to the end of the peak, measured at the baseline. Extrapolation may be required (Figure 1.2).

$w_{0.5}$, *Width of the peak at half height*

The distance, usually in millimeters of chart or in time, from centerline of penmark to centerline of penmark, measured parallel to the baseline at a point halfway between the baseline and the peak maximum.

References

1. Littlewood, A. B., "Gas Chromatography," 2nd ed. Academic Press, New York, 1972.
2. Ettre, L. S., "Open Tubular Columns in Gas Chromatography." Plenum, New York, 1965.
3. International Union of Pure and Applied Chemistry, Commission on Analytical Nomenclature, Recommendations on Nomenclature for Chromatography, *Pure Appl. Chem.* **37,** 445 (1974).

LIQUID PHASES

Relatively few of the many liquid phases available have been successfully coated on glass WCOT columns. Few restrictions are placed on SCOT, PLOT, or whiskered columns (Sections 2.2 and 3.4). The temperature limits are those suggested by the supplier and refer to packed columns. Column life, particularly for glass WCOT columns, will be extended considerably if somewhat lower maximum temperatures are observed (Chapter 10). McReynolds' constants, determined after the concept of Rohrschneider (Section 11.2), were obtained [1, 2] and are reprinted here with the permission of the *Journal of Chromatographic Science* and Supelco, Inc. Remarks include "equivalent" liquid phases and cautionary notes.

Liquid phase	C° min/max	McReynolds' constants					Remarks
		x'	y'	z'	u'	s'	
Alkaterge T							
Amine 220	5/180	117	380	181	293	133	
Apiezon L	50/250	32	22	15	32	42	See SP 21
Armeen SD	30/75						
Bentone 34	0/180						
7,8-Benzoquinoline	55/150						
Benzyl cyanide (phenylacetonitrile)	25/100						
Benzyl cyanide–silver nitrate	0/35						
Benzyldiphenyl	60/120						
n,n-bis(2cyanoethyl) formamide	0/125	690	991	853	1110	1000	
Bis(2-ethoxyethyl) adipate	0/150						
Bis(2-ethylhexyl)tetrachlorphthalate	0/150	112	150	123	168	181	
Bis(2-methoxyethyl) adipate	20/100						
Butanediol adipate	60/225						
Butanediol succinate	50/225	370	571	448	657	611	
Carbowax 400	10/100						
Carbowax 600	30/125						
Carbowax 1000	40/150						
Carbowax 1500	40/175	347	607	418	626	589	
Carbowax 1540	50/175	371	639	453	666	641	
Carbowax 4000	60/200	317	545	378	578	521	
Carbowax 400 monostearate	60/200	282	496	331	517	467	
Carbowax 20 M	60/225	322	536	368	572	510	See also Superox, Section 10.3
Carbowax 20 M–terephthalic acid	60/225	321	537	367	573	520	
Cyclohexanedimethanol succinate	100/250	269	446	328	493	481	
Dibutyl maleate	0/50						
Didecyl phthalate	10/175	136	255	213	320	235	
Diethylene glycol adipate (DEGA)	0/200	378	603	460	665	658	
Diethylene glycol sebacate	80/200						
Diethylene glycol succinate (DEGS)	20/200	496	746	590	837	835	
Di(2-ethylhexyl)sebacate	0/125	72	168	108	180	125	
Diiodecyl phthalate	0/175	84	173	137	218	155	
2,4-Dimethylsulfolane	0/50						
Dinonyl phthalate	20/150	83	183	147	231	159	
Dioctyl sebacate	0/125	72	168	108	180	123	

Liquid phase	C° min/max	McReynolds' constants					Remarks
		x'	y'	z'	u'	s'	
ECNSS-M	30/200	421	690	581	803	732	
EGSS-X	90/200	484	710	585	831	778	
EGSP-Z		308	474	399	548	549	
Emulphor ON-870	0/200	202	395	251	395	344	
Epon 1001	50/225	284	489	406	539	601	
Ethofat 60/25	50/125	191	382	244	380	333	
Ethylene glycol adipate	100/225	372	576	453	655	617	
Ethylene glycol phthalate	100/200	453	697	602	816	872	
Ethylene glycol sebacate	100/200						
Ethylene glycol succinate	100/200	537	787	643	903	889	
Ethylene glycol tetrachlorphthalate	120/200	307	345	318	428	466	
FFAP	50/250	340	580	397	602	625	See SP 1000
Fluorolube GR 362	0/75						
Hallcomid M-18	40/150	079	268	130	222	146	
Hallcomid M-18-OL	8/150	089	280	143	239	165	
Halocarbon 10-25	20/100	047	070	108	133	111	
Halocarbon K-352	0/250	047	070	073	238	146	
Halocarbon wax	50/150	055	071	116	143	123	
n-Hexadecane	18/50						
1,2,3,4,5,6-Hexakis (2-cyanoethoxy-cyclohexane)	125/150	567	825	713	978	901	
Hexamethylphosphoramide	0/50						
Hyprose SP 80	0/175	336	742	492	639	727	
Igepal CO 880	100/200	259	461	311	482	426	
Igepal CO 990	100/200	298	508	345	540	475	
JXR silicone	0/300	015	053	045	064	041	See SP 2100
n-Lauryl-L-valyl-t-butylamide	60/140						
LAC-1-R-296	0/200	377	601	458	663	655	
LAC-2-R-446	50/200	387	616	471	679	667	
LAC-3-R-728	0/200	502	755	597	849	852	
Neopentyl glycol adipate	50/225	234	425	312	402	438	
Neopentyl glycol isophthalate	0/250						
Neopentyl glycol sebacate	50/225	172	327	225	344	326	
Neopentyl glycol succinate	50/225	272	469	366	539	474	
β,β-Oxydipropionitril	0/75						

Liquid phase	C° min/max	McReynolds' constants					Remarks
		x'	y'	z'	u'	s'	
Phenyldiethanolamine succinate	0/200	386	555	472	674	654	
Polyethylene imine	0/175	322	800		573	524	
PPE-20 (poly-M-phenoxylene)	125/375	257	355	348	433		
Tetraethylenepentamine	0/125						
1,2,3,4-Tetrakis-(2-cyanoethoxy)butane	110/200	617	860	773	1048	941	
THEED (tetrahydroxy- ethylenediamine)	0/150	463	942	626		893	
Tricresyl phosphate	20/125	176	321	250	374	299	
Trimer acid	0/150	094	271	163	182	378	
1,2,3-Tris-(2-cyanoethoxy)propane	0/175	594	857	759	1031	917	
Triton X-100	0/200	203	399	268	402	362	
Triton X-305	0/200	262	467	314	388	430	
Tween 80	0/150	227	430	283	438	396	
UCON LB-550-X	0/200	118	271	158	243	206	
UCON 50-HB-280-x	0/200	177	362	227	351	302	
UCON 50-HB-2000	0/200	202	394	253	392	341	
UCON 50-HB-5100	0/200	214	418	278	421	375	
Versamid 900	190/275						
Versamid 930	115/150	109	313	144	211	209	
Xylenyl cyanide (3-cyano-o-xylene)							
Zinc stearate	0/150	061	231	059	098	544	
PPE-21	125/375	232	350	398	413		
Poly-A, Poly-1, Poly-S-179							
Polyphenyl ether (5 rings) OS 124	0/200	176	227	224	306	283	
Polyphenyl ether (6 rings) OS 138	0/225	182	233	228	313	293	
Polypropylene glycol	0/150	128	294	173	264	226	
Polypropylene glycol–silver nitrate	0/50						
Polypropyleneimine	0/200	122	435	168	263	224	
Polyvinylpyrrolidone (PVP)	0/225						
Propylene carbonate	0/50						
Quadrol	0/150	214	571	357	472	489	
Reoplex-400 (polyester)	0/200	364	619	449	647	671	
DC11	0/300	017	086	048	069	056	
DC200, 500 CSTKS	0/200	016	057	045	066	043	

Liquid phase	C° min/max	McReynolds' constants					Remarks
		x'	y'	z'	u'	s'	
DC200, 12, 500 CSTKS	0/250	016	057	045	066	043	
DC550	−20/250	074	116	117	178	135	
DC560 (F60)	0/300	032	072	070	100	068	
DC710	−5/250	107	149	153	228	190	
DC OF 1 (FS 1265)	0/250	144	233	355	463	305	
GE SE 30 GC grade	50/300	015	053	044	064	041	
GE SE 52	50/300	032	072	065	098	067	
GE SF 96	0/250	012	053	042	061	037	
GE XE 60, cyanoethyl	0/250	204	381	340	493	367	
OV 1, methyl gum	100/350	016	055	044	065	042	See SP 2100
OV 3, 10% phenyl	0/350	044	086	081	124	088	
OV 7, 20% phenyl	0/350	069	113	111	171	128	
OV 11, 35% phenyl	0/350	102	142	145	219	178	
OV 17, 50% phenyl	0/375	119	158	162	243	202	See SP 2250
OV 22, 65% phenyl	0/350	160	188	191	283	253	
OV 25, 75% phenyl	0/350	178	204	208	305	280	
OV 101, methyl liquid	0/350	017	057	045	067	043	
OV 210, trifluoropropyl	0/275+	146	238	358	468	310	
OV 225, cyanopropyl phenyl	0/265+	228	369	338	492	386	
OV 275	25/250	629	872	763	1106	849	
Silar 5 CP	0/250	319	495	446	637	531	See SP 2300
Silar 10 CP	0/250	520	757	660	942	800	See SP 2340
SP 400, chlorophenyl	0/350	032	072	070	100	068	
SP 2100, methyl	0/350	017	057	045	067	043	
SP 2250, 50% phenyl	0/375	119	158	162	243	202	
SP 2300, 36% cyanopropyl	25/275+	316	495	446	637	530	See Silar SCP
SP 2310, 55% cyanopropyl	25/275+	440	637	605	840	670	
SP 2330, 68% cyanopropyl	25/275+	490	725	630	913	778	
SP 2340, 75% cyanopropyl	25/275+	520	757	659	942	800	See Silar 10 C
SP 2401, trifluoropropyl	0/275+	146	238	358	468	310	
UC W982	0/250	016	055	045	066	042	See SP 2100
Siponate DS 10	0/200	099	569	320	344	388	
SP 216 PS	25/200	632	875	733	1000	680	
SP 525	60/275	225	255	253	368	320	

Liquid phase	C° min/max	McReynolds' constants					Remarks
		x'	y'	z'	u'	s'	
SP 1000	50/275	332	555	393	583	546	
SP 1200	25/200	067	170	103	203	166	
Span 80	15/150	097	266	170	216	268	
Squalane	20/100	000	000	000	000	000	
Squalene	0/100	152	341	238	329	344	
STAP	100/225	345	586	400	610	627	
Sucrose acetate isobutyrate (SAIB)	0/200	172	330	251	378	295	
Tergitol NPX	10/175	197	386	258	389	351	
Tetracyanoethylated pentaerythritol	30/175	526	782	677	920	837	

References

1. **McReynolds, W. O.**, *J. Chromatogr. Sci.* **8**, 685 (1970).
2. Supelco, Inc., Bellefonte, Pennsylvania 16823. Catalog No. 10 (1978).

POROUS POLYMER DATA

These materials, although occasionally of great value to specific gas chromatographic separations, find their greatest utility in sample concentration (Section 12.10). Additional data on the relative retentions of a number of solutes on several different porous polymers were published by Lindsay Smith *et al.* in the *Journal of Chromatography* (**151**, p. 27, 1978).

Porapaks

Characterized by short retentions for water, alcohols, and glycols, the Porapaks are among the most versatile of the porous polymers. They are assigned a polarity (N, P, Q, R, S, T) based on an increasing affinity for water. Porapak P is a tripolymer of styrene, ethyl vinylbenzene, and divinylbenzene; Porapak Q is a copolymer of ethyl vinylbenzene and divinylbenzene; Porapak R is N-vinylpyrrolidine; and Porapak T is ethylene glycol dimethacrylate. All are available in mesh ranges of 50–80, 80–100, 100–120, 150–200, and 200–325. According to the supplier, the maximum temperature limit is 250°C, except for N and T, which have a suggested maximum temperature limit of 190°C: even at lower temperatures, considerable degradation can be observed in the presence of oxygen. Complete information can be obtained by requesting Product Brochure PB-71-205 and Data Sheet DS-71-004 from the supplier, Waters Associates, Inc., 61 Fountain Street, Framingham, Massachusetts 01701.

The following data, furnished by the supplier, permit some estimation of their various efficiencies in trapping various types of volatile constituents. The data show the elution time in minutes for various substances on different Porapaks, contained in a 6-ft column and utilizing helium carrier gas.

Tenax GC

A porous polymer based on 2,6-diphenyl-p-phenylene oxide, marketed by Enka N.V. of The Netherlands, and distributed in the United States by Applied Science Laboratories, Inc., P.O. Box 440, State College, Pennsylvania 16801. Although the capacity is lower than that of some of the other porous polymers, its excellent thermal stability makes it a very useful material (see Sections 12.10, 15.2, 15.7, 15.12). It is available in 60–80 and 35–60 mesh sizes. When used for chromatographic purposes, the suppliers caution that the mixture to be separated should be injected several times prior to qualitative or quantitative analysis. This hints at a "demand capacity" and suggests rigorous checks of trapping and recover efficiencies may be in order. Full details are to be found in Technical Bulletin No. 24, available from Applied Science Laboratories. The following data are excerpted from that bulletin.

Chromosorbs, Century Series

Not to be confused with the normal diatomecious earth-type Chromosorbs, these porous polymers supplied by Johns-Manville, Celite Division, 22 East 40th Street, New York, New York 10016 are the subject of Technical Bulletin FF-202A, obtainable from the supplier. They are available in mesh sizes of 50–60, 60–80, 80–100, and 100–120.

Retention Times of Selected Compounds on Porous Polymer Columns

Porapak type	P	Q	R	S	T
Column Temperature 0°C					
Air	0.39	0.30	0.35	0.33	0.44
Methane	0.49	0.47	0.61	0.8	0.66
Carbon dioxide	0.83	1.1	1.72	1.7	3.6
Ethylene	1.27	2.3	3.2	4.0	3.8
Acetylene	1.46	2.3	4.8	5.0	11.0
Ethane	1.76	3.3	4.35	5.6	4.5
Water	4.88	5.9	50.0	33.2	110
Propylene	5.0	16	18.7	26	27
Propane	7.9	21	22.8	29	24
Methyl chloride	8.6	19	—	34	50
Column Temperature 32°C					
Air	0.31	0.29	0.23	0.40	0.42
Methane	0.39	0.59	0.33	0.52	0.45
Carbon dioxide	—	—	0.56	0.78	0.85
Ethylene	0.54	0.88	0.75	1.1	0.95
Acetylene	—	0.88	0.95	1.2	1.4
Ethane	0.54	0.88	0.95	1.3	1.0
Water	1.00	1.2	3.5	3.2	7.1
Propylene	—	—	2.3	3.6	2.6
Propane	1.00	2.9	2.6	3.6	2.6
Methyl chloride	1.2	3.2	2.8	4.5	4.2
Vinyl chloride	1.6	5.6	4.7	7.1	6.8
Methyl alcohol	1.4	2.9	5.6	7.8	10.0
Ethylene oxide	—	—	4.2	7.8	9.5
Ethyl chloride	2.7	9.7	8.5	13.5	13
Acetone	5.3	17	15	36	41
Ethyl alcohol	2.8	8.2	14	19.5	26
Pentane	6.7	—	20	36	22
Cyclopentane	10.0	—	25	49	28
Butadiene	2.4	7.9	7.5	11	9.1
Isopropyl alcohol	5.2	18	27	40	54
Acetonitrile	4.9	12	13	20	41
Acrylonitrile	7.0	21	21	32	58
Diethyl ether	6.6	30	22	35	28
Methylene chloride	5.5	19	19	29	34
n-Propyl alcohol	7.7	29	47	65	85
t-Butyl alcohol	7.7	39	55	74	98

Retention Times of Selected Compounds on the Porapaks[a]

Porapak type	P	Q	R	S	T
Air	0.28	0.3	0.3	0.39	0.32
Methane	—	0.35	0.39	0.46	0.40
Carbon dioxide	—	—	0.39	0.62	0.50
Ethylene	—	0.5	0.52	0.70	0.53
Acetylene	—	—	0.58	0.62	0.64
Ethane	—	0.55	0.58	0.70	0.53
Water	0.48	0.65	1.0	1.4	1.7
Propylene	—	—	0.91	1.4	0.91
Propane	—	1.16	0.91	1.4	0.91
Methyl chloride	—	1.21	1.1	1.6	1.3
Vinyl chloride	—	1.7	1.6	2.2	1.6
Methyl alcohol	0.54	1.1	1.5	1.9	2.2
Ethylene oxide	—	—	1.35	—	2.1
Ethyl chloride	—	2.5	2.3	3.1	2.7
Acetone	1.00	3.5	3.4	4.5	5.8
Ethyl alcohol	0.75	2.2	2.8	3.4	4.1
Pentane	1.11	5.6	4.5	5.8	3.6
Cyclopentane	1.3	7.1	5.5	7.5	4.5
Butadiene	0.73	—	2.0	2.7	1.9
Isopropyl alcohol	0.94	3.8	4.5	5.7	6.7
Acetonitrile	1.09	3.2	3.2	4.2	6.2
Acrylonitrile	1.22	4.1	4.4	5.7	7.7
Diethyl ether	0.93	4.9	4.2	5.5	4.3
Methylene chloride	1.2	4.2	3.7	5.3	5.0
n-Propyl alcohol	1.2	5.0	6.5	7.8	9.5
t-Butyl alcohol	1.25	6.1	7.2	8.2	10.5
Hexane	2.1	13.5	8.5	13	8.1
Cyclohexane	2.35	17	10.5	17	10.0
Benzene	2.65	14.5	10.0	16	13.0
Nitro ethane	2.55	9.7	8.5	13	22
Ethylene glycol	3.1	10.5	23	28	53
Propylene glycol	4.0	18.5	34	42	69
Formic acid	—	2.05	—	26	13.5
Acetic acid	—	4.3	14.5	26	21
Propionic acid	—	10.5	32	—	46

[a] 6-ft × ¼-in. Column at 157°C.

Properties of the Chromosorb "Century Series" Porous Polymers

Physical properties	101	102	103	104	105
Type	STY–DVB	STY–DVB	Cross-linked polystyrene	ACN–DVB	Polyaromatic
Free fall density (g/cc)	0.30	0.29	0.32	0.32	0.34
Surface area (m²/g)	Less than 50	300–400	15–25	100–200	600–700
Average pore diameter (μ)	0.3–0.4	0.0085	0.3–0.4	0.06–0.08	0.04–0.06
Water affinity	Hydrophobic	Hydrophobic	Hydrophobic	Hydrophobic	Hydrophobic
Color	White	White	White	White[a]	White
Temperature limit					
(isothermal) (°C)	275	250	275	250	250
(programmed) (°C)	325	300	300	275	275

[a]Turns yellow after column conditioning and brown after extended use, but this does not affect column performance.

Retention Times and Retention Indices on Chromosorb "Century Series"[a]

Compound	b.p. (°C)	Molecular weight	101		102	
			Retention time (min)	Retention index	Retention time (min)	Retention index
Alcohols						
n-C_1	64.7	32.00	0.46	440	0.43	360
n-C_2	78.4	46.10	0.57	495	0.65	425
n-C_3	97.2	60.10	0.83	595	1.11	510
n-C_4	118.0	74.12	1.31	700	2.08	615
n-C_5	119.9	88.15	2.13	805	3.91	725
iso-C_3	82.4	60.10	0.71	555	0.94	485
iso-C_4	108.1	74.12	1.12	665	1.84	595
iso-C_5	130.0	88.15	1.87	775	3.55	700
sec-C_4	99.5	74.12	0.98	635	1.60	570
t-C_4	82.6	74.12	0.74	565	1.20	525
t-C_5	102.4	88.15	1.24	690	2.38	635
Ketones						
Acetone	56.5	50.10	0.70	555	0.91	480
MEK	79.6	72.10	1.03	645	1.61	570
DiEK	102.0	86.10	1.60	745	2.88	665
Ethers						
Dipropyl	90.1	102.18	1.27	645	0.81	510
Dibutyl	142.4	130.23	3.26	885	5.21	880
Chloroalkyls						
CH_2Cl_2	40.2	84.93	0.77	590	1.10	510
$CHCl_3$	61.2	119.40	1.07	675	1.82	590
CCl_4	76.8	153.82	1.27	720	2.58	650
1,2-Cl_2Et	83.5	98.96	1.32	725	2.15	620
1,2-Cl_2Pr	96.4	113.00	1.62	775	3.20	685
Glycols						
1,2-C_2	196.0	62.10	1.91	780	1.40	625
1,3-C_3	97/6 mm	76.10	3.67	910	2.61	745
1,4-C_4	228.0	90.12	6.63	1030	4.78	895
1,3-C_4	116/20 mm	90.12	4.47	950	3.55	810
2,3-C_4	—	90.12	2.83	865	2.54	740
Butene	—	—	7.30	1050	4.88	865
Butyne	—	—	9.01	1090	5.41	885
Aldehydes						
C_2	20.20	44.05	0.48	440	0.53	400
n-C_3	49.50	58.08	0.69	550	0.90	475
n-C_4	75.70	72.11	1.02	645	1.57	565

103		104		105	
Retention time (min)	Retention index	Retention time (min)	Retention index	Retention time (min)	Retention index
0.46	420	1.05	625	0.50	365
0.57	495	1.38	690	7.79	435
0.87	595	2.28	795	1.51	535
1.41	705	3.99	905	3.07	655
2.36	810	5.88	980	6.30	760
0.67	540	1.57	720	1.17	490
1.19	660	3.19	865	2.61	625
2.07	780	5.96	985	5.51	755
1.09	655	2.57	820	2.26	615
0.77	575	1.96	765	1.64	545
1.34	695	3.43	780	3.44	665
0.43	530	1.85	755	1.06	465
0.67	640	2.92	850	2.03	580
1.04	735	4.54	935	3.06	675
0.84	690	—	—	—	—
2.23	885	—	—	—	—
0.51	575	1.54	715	—	—
0.74	660	—	—	—	—
0.92	710	1.66	730	2.91	635
0.91	705	—	—	—	—
1.13	755	—	—	—	—
—	—	—	—	—	—
—	—	—	—	—	—
—	—	—	—	—	—
—	—	—	—	—	—
—	—	—	—	—	—
—	—	—	—	—	—
—	—	—	—	—	—
0.45	410	1.00	615	0.59	395
0.66	535	1.60	725	1.05	480
0.99	630	2.65	830	2.02	580

Compound	b.p. (°C)	Molecular weight	101		102	
			Retention time (min)	Retention index	Retention time (min)	Retention index
Acids						
C$_2$	118.1	60.50	0.88	610	0.70	435
C$_3$	141.1	74.10	1.37	715	1.07	495
C$_4$	164.0	88.10	2.15	805	2.00	610
C$_5$	186.4	102.13	3.60	910	3.60	705
Acetates						
Methyl	57.3	74.08	0.75	570	1.08	505
Ethyl	77.1	88.10	1.01	655	1.70	580
Propyl	101.6	102.13	1.55	735	3.10	680
Butyl	126.1	116.16	2.53	540	5.77	780
Nitriles						
Aceto	81.6	41.05	0.77	580	0.87	460
Propio	97.2	55.08	1.07	660	1.39	550
Amines						
n-C$_1$	−6.7/755 mm	31.06	—	—	—	—
n-C$_2$	16.6	45.08	—	—	—	—
n-C$_3$	49.0	59.11	—	—	—	—
n-C$_4$	77.8	73.14	—	—	—	—
n-C$_5$	103.0	87.17	—	—	—	—
n-C$_6$	129.0	101.19	—	—	—	—
iso-C$_3$	32.0	59.11	—	—	—	—
iso-C$_4$	66.0	73.14	—	—	—	—
iso-C$_5$	95.00	87.17	—	—	—	—
t-C$_4$	45.00	73.14	—	—	—	—
t-C$_5$	76.00	87.17	—	—	—	—
Dimethyl	6.9	45.09	—	—	—	—
Diethyl	56	73.14	—	—	—	—
Triethyl	89	101.19	—	—	—	—
Dipropyl	109	101.19	—	—	—	—
Diisopropyl	83	101.19	—	—	—	—
Diamines						
1,2-C$_2$	117.2	60.10	—	—	—	—
1,3-C$_3$	138.0	74.13	—	—	—	—
1,4-C$_4$	54/11 mm	88.15	—	—	—	—
1,5-C$_5$	—	102.18	—	—	—	—
1,2-C$_3$	—	74.13	—	—	—	—
Anilines						
Aniline	184.4	91.13	—	—	—	—
n-Methyl	81/14 mm	107.16	—	—	—	—

103		104		105	
Retention time (min)	Retention index	Retention time (min)	Retention index	Retention time (min)	Retention index
—	—	6.02	985	1.74	555
—	—	9.32	1070	3.46	655
—	—	15.02	1155	6.91	765
—	—	25.92	1255	14.40	875
0.48	560	1.51	710	1.20	500
0.64	630	2.14	785	2.09	585
1.00	730	3.85	885	4.09	685
1.66	825	6.06	990	8.24	790
0.75	565	3.05	855	1.01	480
1.04	640	4.37	925	1.73	560
0.43	390	—	—	—	—
0.53	470	—	—	—	—
0.77	575	—	—	—	—
1.21	670	—	—	—	—
1.98	770	—	—	—	—
3.34	865	—	—	—	—
0.63	525	—	—	—	—
1.05	645	—	—	—	—
1.76	745	—	—	—	—
0.73	560	—	—	—	—
1.26	680	—	—	—	—
0.45	410	—	—	—	—
0.83	590	—	—	—	—
1.45	710	—	—	—	—
2.00	770	—	—	—	—
1.26	680	—	—	—	—
1.69	740	—	—	—	—
2.90	840	—	—	—	—
5.06	940	—	—	—	—
8.82	1040	—	—	—	—
2.11	785	—	—	—	—
2.97	1140	—	—	—	—
3.92	1210	—	—	—	—

| Compound | b.p. (°C) | Molecular weight | 101 | | 102 | |
			Retention time (min)	Retention index	Retention time (min)	Retention index
n-Ethyl	97/21 mm	121.18	—	—	—	—
n-Propyl	85/5 mm	135.2	—	—	—	—
n-*n*-Dimethyl	194.2	121.18	—	—	—	—
Aryls						
Benzene	80.1	78.10	1.60	745	2.61	650
Methylbenzene	110.6	92.14	2.59	845	4.81	750
Ethylbenzene	136.2	106.17	4.10	935	8.63	845
Miscellaneous						
Cyclohexane	80.7	84.16	1.37	700	2.62	560
Cyclohexene	83.3	82.15	1.51	730	2.71	655
Dioxane	—	—	1.84	775	3.00	675
Nitromethane	65	61.04	1.05	650	1.12	510
Phenol	182	94.11	8.13	1070	6.01	905
Pyridine	115.5	79.10	2.28	850	3.68	705

103		104		105	
Retention time (min)	Retention index	Retention time (min)	Retention index	Retention time (min)	Retention index
4.61	1245	—	—	—	—
6.47	1335	—	—	—	—
4.06	1220	—	—	—	—
1.55	720	2.72	835	2.93	635
2.57	820	4.56	935	—	—
4.24	910	7.05	1020	—	—
1.38	695	1.50	710	3.02	640
1.51	715	1.75	735	—	—
1.23	770	4.55	935	—	—
—	—	4.62	935	1.49	530
—	—	—	—	—	—
1.60	820	7.38	1025	5.09	720

[a] 4-ft × 3-mm i.d. glass column, 80–100 mesh, 190°C, 60 ml/min He.

Some Compounds That Have Been Separated on Tenax-GC[a]

	Elution temperature (°C)	Retention time (min)
Ethoxylated lauryl alcohols		
Technical lauryl alcohol		
+5 ethylene oxide units	200–380	50
Alcohols		
1-Decanol	260	1.5
1-Dodecanol	265	2.5
1-Hexadecanol	290	7
1-Octadecanol	310	10
Polyethylene glycols		
4 Ethylene oxide units	260	7.5
8 Ethylene oxide units	350	18.5
Diols		
Ethylene glycol	155	0.5
1,3-Propanediol	160	2
1,4-Butanediol	175	4.5
1,5-Pentanediol	190	7
1,6-Hexanediol	205	9.5
1,7-Heptanediol	225	12
1,8-Octanediol	240	15
1,9-Nonanediol	255	17.5
1,10-Decanediol	270	20
Phenols		
Phenol	210	2.5
m-Cresol	220	4
2,6-Dimethylphenol	225	5
3,5-Dimethylphenol	230	6
o-Aminophenol	245	9
m-Aminophenol	260	11.5
m-N,N-Diethylaminophenol	280	15.5
Naphthol	285	17
p-Phenylphenol	320	23.5
Methyl esters of dicarboxylic acids		
Hexanedioic acid	240	5
Heptanedioic acid	250	6.5
Octanedioic acid	265	8
Nonanedioic acid	275	9.5
Decanedioic acid	290	11
Dodecanedioic acid	315	14.5
Ethanolamines		
Ethanolamine	155	0.5
Diethanolamine	210	7.5

	Elution temperature (°C)	Retention time (min)
Triethanolamine	270	15
Alkyl amines		
Propylamine	130	5
1-Methylpropylamine	140	7
Butylamine	155	9
3-Methylbutylamine	170	11.5
Heyxlamine	195	16
Diamines		
1,5-Pentanediamine	190	2.5
1,6-Hexanediamine	195	4
1,7-Heptanediamine	210	7
1,8-Octanediamine	220	10
1,9-Nonanediamine	230	13
1,10-Decanediamine	245	16
1,12-Dodecanediamine	265	21.5
Amides		
Formamide	130	3
Acetamide	140	6
N,N-Dimethylformamide	140	7
N-Methylacetamide	145	9
Propionamide	150	10.5
N,N-Dimethylacetamide	155	11.5
Aromatic amines		
Pyridine	160	1.5
Morpholine	165	2.5
Aniline	190	6.5
o-Toluidine	205	9
Carbonyl compounds		
Acetone	120	2.5
Glycolaldehyde	130	4
2,4-Pentanedione	170	9
Cyclohexanone	190	11.5
Benzaldehyde	210	13.5
5-Nonanone	200	9.5
6-Undecanone	225	15
3-Dodecanone	240	18.5
4-Tridecanone	255	21
5-Tetradecanone	270	23.5
6-Pentadecanone	280	26
6-Hexadecanone	290	28.5

[a]The elution temperatures and retention times are derived from chromatograms obtained with temperature programming.

SILYLATION AND DERIVATIZATION REACTIONS

The word "silylation" usually means "trimethylsilylation," the substitution of an —Si(CH₃)₃ group for an active hydrogen, but it is also sometimes used to designate the substitution or attachment of such organo-silicon groups as dimethylsilyl, —SiH(CH₃)₂, or chloromethyldimethylsilyl, Si(CH₃)₂CH₂Cl. Reagents, reaction kits (*vide infra*) and an invaluable monograph, "Handbook of Silylation" [1], are available from Pierce Chemical Company, P.O. Box 117, Rockford, Illinois 61105. An authoritative test and reference on silylation and silylation reactions is also available [2].

Water will decompose both the trimethylsilyl (TMS) reagents and their derivatives; syringes, reaction vessels, reagents, and sample should be dry. The hydrolysis product is hexamethyldisiloxane, (Ch₃)₃SiOSi(Ch₃)₃, and an increase in this peak can be used to check the moisture content of sample or reagents. Reaction time may vary from a few seconds to several hours. To establish the minimum necessary time, samples should be injected immediately after mixing and at 5-, 15-, and 30-min, and 1-, 4-, and 8-hr intervals. The shortest time after which there has been no increase in product peak(s) for at least two subsequent analyses may be taken as the necessary reaction time [1].

Vaporization of the derivatives on a metal surface may cause decomposition; glass inlet systems are strongly recommended. Trace amounts of moisture in the carrier gas may cause decomposition of derivatives; sugar derivatives are unusually stable, and amino acid derivatives, unusually labile to trace levels of moisture. Some authorities recommend oxygen and moisture scavengers in the carrier gas line.

Solvents must be anhydrous; pyridine is most generally used, and some feel that its ability to accept HCl gives it a strong advantage [1, 2]. Dimethylformamide has been widely used, and dimethylsulfoxide and tetrahydrofuran are of value in some special cases. Acetonitrile, in spite of its high toxicity in both the liquid and vapor forms, has seen some limited use.

Because the reagent is used in excess, the injected sample contains appreciable amounts of unreacted silylating reagent. This precludes the use of liquid phases with reactive hydrogens; the silicones (Appendix II) have been most widely used, although the Apiezons have also proven useful.

Reagents

A few of the more popular silylating reagents follow. These, and a wide variety of others, are readily available from several sources, including Pierce Chemical Company.

BSA: *N,O*-bis(trimethylsilyl) acetamide

BSTFA: *N,O*-bis(trimethylsilyl trifluoroacetamide

Advantage is that the by-products of its reaction exhibit lower boiling points. Supplied pure, or with 1% TMCS (*vide infra*) as a catalyst.

HMDS: Hexamethyldisilazane

Somewhat slower acting, but may be catalyzed with TMCS (*vide infra*).

TMCS: Trimethylchlorosilane

Somewhat slow when used alone, a powerful catalyst in combinations.

Silylation Kits

Available in individual "one-shot" sealed ampoules or in septum-sealed reagent bottles, these contain solvent, reagent, and (where required) catalyst in a ready-to-use form. A representative sampling would include the following.

TRI-SIL

Used for a broad range of compounds containing hydroxyl, carboxyl, and amine groups.

TRI-SILZ

Will react in the presence of moderate amounts of water, particularly useful in sugar derivatization.

TRI-SIL-BSA

A powerful and general silyl donor, available in pyridine (formula P) or dimethylformamide (formula D).

Typical Reactions

A few general and classical reactions follow. A detailed survey of specific derivatization reactions is to be found in Cram and Juvet [3].

General

5-10-mg samples plus 1 cm^3 TRI-SIL are shaken in a plastic stoppered vial ~30 sec. If solution is difficult, warm to 75°-85°C. Wait 5 min and inject [4].

Sugars, etc. (see [5])

Up to 10 mg of carbohydrate in 1-cm^3 anhydrous pyridine is treated with 0.2-cm^3 HMDS and 0.1-cm^3 TMCS. Shake until dissolved (30 sec); warming may be required as above. After 5 min, begin injection series to determine reaction time.

Syrups, etc.

To 60-70-mg heavy syrup add 1-cm^3 pyridine, 0.9-cm^3 HMDS, and 0.1-cm^3 trifluoroacetic acid. Shake 30 sec, let stand 15 min, and begin injection series. Sugars, etc.: 5-20-mg sample is added to 0.5-cm³ pyridine containing 12-mg methoxylamine·HCl and heated to 80°C for 2 hr; 0.3-cm³ BSTFA containing 4% TMCS is added, the mixture is heated 15 min at 80°C, and it is then injected [6].

Steroids (see [7, 8])

0.1–5-mg sample is added to 0.2–0.4-cm³ BSA; if necessary to effect solution, 0.1–0.2-cm³ pyridine and heat to 60°C may be used. Reaction time varies from a few minutes to several hours; largely restricted to sterically unhindered positions [6]. For moderately hindered positions (e.g., 11-β-ols) plus unhindered groups, substitute TRI-SIL'BT' for BSA. Complete reaction of the moderately hindered sites will require 6–24 hr at room temperature, or 3–6 hr at 60°C [6]. To silylate even the most hindered groups, substitute TRI-SIL'TBT' for BSA. Complete reaction usually requires 6–24 hr at 60°–80°C [7]. Ketosteroids have been analyzed as *o*-pentafluorobenzyloxime derivatives [9] by a technique that was later adapted to the analysis of dehydroepiandresterone in human plasma [10].

Amino acids

A variety of methods have been utilized to prepare volatile derivatives of amino acids, suitable for gas chromatographic analysis. Horning *et al.* [11] developed a method that combines silylation of hydroxyl groups with acylation of amino groups. It depends on the fact that *N*-trimethylsilylimidazole (TSIM) silylates only the hydroxyl groups, blocking these from further reaction but leaving the amino exposed for acylation to produce the volatile *O*-trimethylsilyl, *N*-fluoroacyl derivatives.

To about 1 mg of sample dissolved in 0.1-cm³ acetonitrile is added 0.2-cm³ TSIM, and the sealed vial is heated 3 hr at 60°C. Depending on which derivative is desired, 0.1 cm³ of either *N*-heptafluorobutyrylimidazole (HFBI) or *N*-trifluoroacetylimidazole (TFAI) is added, and reheated 30 min at 60°C. Aliquots of the reaction mixture are injected directly. The inner esters of acyl amino acids, azlactones, or oxazolin-5-ones have been suggested as alternatives to the *N*-acyl esters as volatile derivatives for gas chromatography [12]. Mixtures of α-methyl amino acids have also been analyzed as their corresponding 2-phenyl-oxazolin-5-ones, obtained by *N*-benzolyllation of the amino group followed by formation of the inner ester with dicylclohexyl carbodiimide [13].

Thenot and Horning [14] later reported that it was possible to use a single reagent, N,N-dimethyl formamide dimethylacetal, to derivatize simultaneously both the carboxyl and amine groups. The former is converted to the methyl ester, and the latter to the N-dimethylaminomethylene derivative; the volatile product is reported to have excellent gas chromatographic and mass spectral characteristics.

The sample is mixed with the reagent, heated 10 min at 60°C, and aliquots of the warm reaction mixture are injected. Scoggins [15] found the method also applicable to diastereomeric diamines.

Several other derivatization techniques for amino acids are to be found in Section 15.3.

Pesticides

The major problem in pesticide analysis is one of clean-up; several recent issues of the *Journal of Chromatographic Science* were devoted to a state-of-the-art review of pesticide analysis (**13**, May–July, 1975).

Fatty acids

A great deal of information is available on the derivatization of fatty acids. Again, recent issues of the *Journal of Chromatographic Science* (**13**, September, October, 1975) were devoted to a state-of-the-art review on fatty acid analysis. Most of the methods in current use, including those cited in Section 15.6, are patterned after the following.

The borontrifluoride (BRF) method is one of the most widely used reagents because of its stability, and the simplicity and rapidity of the reaction [16]. BIF-butanol is also a very popular reagent [17]. In either case, about 25 mg of the fatty acid sample, usually dissolved in 2-cm³ benzene, is mixed with 2 cm³ of reagent and boiled on a steam bath for 3 min. About 1 cm³ of water is added, and the mixture allowed to separate (centrifuge if necessary). The upper benzene layer, containing the fatty acid esters, is used directly for analysis. Methanolic HCl, prepared by bubbling dry HCl gas into dry methanol, has also been used to prepare fatty acid methyl esters and is reported to be particularly useful for the more volatile or short-chain fatty acids.

Thenot *et al.* [18] used dimethylformamide dialkyl acetals to prepare methyl, ethyl, *n*-propyl, *n*-butyl, and *t*-butyl esters of long-chain fatty acids. Readily soluble samples were heated 10 min at 60°C and injected; samples that did not dissolve readily in the reagent were heated in the same way using a 1:1 mixture of reagent and pyridine.

Transesterification of the fatty acids of fats and oils has been accomplished with dry methanolic HCl [19], in one case combined with 2,2-dimethoxypropane to drive the reaction and eliminate the need for elevated temperatures [20]. A base-catalyzed technique, using sodium and potassium methoxide for the quantitative preparation of methyl esters of glycerides, cholesteryl esters, and phospholipids, has also been reported [21].

References

1. "Handbook of Silylation." Pierce Chem. Co., Rockford, Illinois, 1978.
2. Pierce, A. E., "Silylation of Organic Compounds." Pierce Chem. Co., Rockford, Illinois, 1968.
3. Cram, S. P., and Juvet, R. S., *Anal. Chem.* **48**, 411R (1976).
4. Sweeley, C. C., Bentley, R., Makita, M., and Wells, W. W., *J. Am. Chem. Soc.* **85**, 2497 (1963).
5. Laine, R. A., and Sweeley, C. C., *Carbohydr. Res.* **27**, 199 (1973).
6. Chambaz, E. M., and Horning, E. C., *Anal. Lett.* **1**, 201 (1968).
7. German, A., and Horning, E. C., *J. Chromatogr. Sci.* **11**, 76 (1973).
8. Sandra, P., Verzele, M., and van Luchene, E., *Chromatographia* **8**, 499 (1975).
9. Koshy, K. T., Kaiser, D. G., and van der Slik, A. L., *J. Chromatogr. Sci.* **13**, 97 (1975).
10. Nambara, T., Kigasawa, K., Iwata, T., and Ibuki, M. J., *Chromatogr.* **114**, 81 (1975).
11. Horning, M. G., Moss, A. M., Boucher, E. A., and Horning, E. C., *Anal. Lett.* **1**, 311 (1968).
12. Grahl-Nielsen, O., and Solheim, E., *Chem. Commun.* p. 1093 (1972).
13. Grahl-Nielsen, O., and Solheim, E., *Anal. Chem.* **47**, 333 (1975).
14. Thenot, J. P., and Horning, E. C., *Anal. Lett.* **5**, 519 (1972).
15. Scoggins, M. W., *J. Chromatogr. Sci.* **13**, 146 (1975).
16. Metcalfe, L. D., and Schmitz, A. A., *Anal. Chem.* **33**, 363 (1961).
17. Jones, E. P., and Davison, V. L., *J. Am. Oil Chem. Soc.* **62**, 121 (1965).
18. Thenot, J. P., Horning, M. G., Stafford, M., and Horning, E. C., *Anal. Lett.* **5**, 217 (1972).
19. Stoffel, W., Chu, F., and Ahrens, E. H., *Anal. Chem.* **31**, 307 (1959).
20. Mason, M. E., and Waller, G. R., *Anal. Chem.* **36**, 583 (1964).
21. Luddy, E. F., Barford, R. A., and Riemenschneider, R. W., *J. Am. Oil Chem. Soc.* **45**, 549 (1968).

INDEX